PURE AND APPLIED MATHEMATICS

A Program of Monographs, Textbooks, and Lecture Notes

LECTURE NOTES
IN PURE AND APPLIED MATHEMATICS

STABILITY OF DYNAMICAL SYSTEMS

Theory and Applications

Proceedings of the National Science Foundation—Conference Board of Mathematical Sciences Regional Conference held at Mississippi State University

Edited by
JOHN R. GRAEF
Department of Mathematics
Mississippi State University
Mississippi State, Mississippi

MARCEL DEKKER, INC. New York and Basel

Library of Congress Cataloging in Publication Data

Main entry under title:

Stability of dynamical systems.

(Lecture notes in pure and applied mathematics ; 28)
Includes index.
1. Differential equations--Congresses.
2. Integral equations--Congresses. 3. Stability--Congresses. I. Graef, John R., 1942-
II. United States. National Science Foundation.
III. Conference Board of the Mathematical Sciences.
QA371.S825 515'.35 75-40766
ISBN 0-8247-6410-2

MARCEL DEKKER, INC.

270 Madison Avenue, New York, New York 10016

Current printing (last digit):
10 9 8 7 6 5 4 3 2 1

PRINTED IN THE UNITED STATES OF AMERICA

PREFACE

This volume contains the invited addresses and contributed papers presented at a Regional Research Conference sponsored by the National Science Foundation and the Conference Board of the Mathematical Sciences and held at Mississippi State University from August 11 to 15, 1975. The subject of the conference was the stability of dynamical systems, and Professor J. P. LaSalle of the Lefschetz Center for Dynamical Systems, Brown University, was the Principal Lecturer. Professor LaSalle's lecture notes appear as Volume 25 in the SIAM Regional Conference Series in Applied Mathematics.

With the exception of Chapter 1 by Artstein, all the chapters in this volume were presented at the Conference; Artstein's lectures at the Conference appear as an appendix to LaSalle's notes. The contribution by Artstein in this volume is research completed after he returned to the Weizmann Institute, but due to its relevance to the theme of the Conference, it is included here for completeness.

The range of topics discussed in this book -- control theory, mathematical economics, nonlinear oscillations, random integral equations, various stability criteria -- attest to the importance and broad influence that stability theory in general, and LaSalle's famous invariance principle in particular, have had on the mathematical sciences.

I wish to especially thank the National Science Foundation and the Conference Board of the Mathematical Sciences for funding the Conference. I also wish to thank those who gave the invited addresses: Z. Artstein, T. A. Burton, A. G. Kartsatos, and R. Reissig. I am grateful to Paul W. Spikes who helped with the local arrangements for the conference, Gail Hudson who did an excellent job of typing the manuscript, and the staff at Marcel Dekker for their help in the preparation of this volume. A very special thanks goes to J. P. LaSalle not only for serving as Principal Lecturer for the Conference, but also for his many contributions to mathematics which have so strongly influenced current research in stabiliby theory and its applications.

John R. Graef

CONTENTS

CONTRIBUTORS

ZVI ARTSTEIN*
Division of Applied Mathematics, Brown University, Providence, Rhode Island

PREM N. BAJAJ
Mathematics Department, Wichita State University, Wichita, Kansas

T. A. BURTON
Mathematics Department, Southern Illinois University at Carbondale, Carbondale, Illinois

KUANG-HO CHEN
Department of Mathematics, University of New Orleans, Lake Front, New Orleans, Louisiana

ETHELBERT N. CHUKWU
Department of Mathematics, Cleveland State University, Cleveland, Ohio

JOHN R. GRAEF
Department of Mathematics, Mississippi State University, Mississippi State, Mississippi

CHARLES R. JOHNSON
Institute for Physical Science and Technology, University of Maryland, College Park, Maryland

A. G. KARTSATOS
Department of Mathematics, University of South Florida, Tampa, Florida

WADI MAHFOUD†
Mathematics Department, Southern Illinois University at Carbondale, Carbondale, Illinois

---

*Present Address: Department of Mathematics, The Weizmann Institute of Science, Rehovot, Israel

†Present Address: Department of Mathematics, Murray State University, Murray, Kentucky

ROGER McCANN
Department of Mathematics and Statistics, Case Western Reserve University, Cleveland, Ohio

ALEXANDER MORGAN
Mathematics Department, University of Miami, Coral Gables, Florida

K. S. NARENDRA
Department of Engineering and Applied Science, Yale University, New Haven, Connecticut

DEAN A. NEUMANN
Department of Mathematics, Bowling Green State University, Bowling Green, Ohio

R. REISSIG
Institut für Mathematik, Ruhr-Universität Bochum, Federal Republic of Germany

JAMES A. RENEKE
Department of Mathematical Sciences, Clemson University, Clemson, South Carolina

A. S. C. SINHA
Departments of Electrical Engineering and Mathematics, Indiana Institute of Technology, Fort Wayne, Indiana

PAUL W. SPIKES
Department of Mathematics, Mississippi State University, Mississippi State, Mississippi

JAMES R. WARD*
Department of Mathematics, University of Oklahoma, Norman, Oklahoma

GLENN WEBB
Department of Mathematics, Vanderbilt University, Nashville, Tennessee

*Present Address: Department of Mathematics, Pan American University, Edinburg, Texas

PARTICIPANTS

O. R. AINSWORTH, Mathematics Department, University of Alabama, University, Alabama

ZVI ARTSTEIN, Division of Applied Mathematics, Brown University, Providence, Rhode Island

T. A. ATCHISON, Department of Mathematics, Mississippi State University, Mississippi State, Mississippi

PREM N. BAJAJ, Mathematics Department, Wichita State University, Wichita, Kansas

RAVI BAJAJ, Wichita, Kansas

JOHN S. BRADLEY, Mathematics Department, University of Tennessee, Knoxville, Tennessee

L. P. BURTON, Department of Mathematics, Auburn University, Auburn, Alabama

T. A. BURTON, Mathematics Department, Southern Illinois University at Carbondale, Carbondale, Illinois

KUANG-HO CHEN, Department of Mathematics, University of New Orleans, Lake Front, New Orleans, Louisiana

A. K. CHOUDHURY, Department of Electrical Engineering, Howard University, Washington, D.C.

ETHELBERT N. CHUKWU, Department of Mathematics, Cleveland State University, Cleveland, Ohio

ROBERT E. FENNELL, Department of Mathematical Sciences, Clemson University, Clemson, South Carolina

WILLIAM E. FITZGIBBON, Department of Mathematics, University of Houston, Houston, Texas

JOHN R. GRAEF, Department of Mathematics, Mississippi State University, Mississippi State, Mississippi

LOUIS J. GRIMM, Department of Mathematics, University of Missouri, Rolla, Missouri

JOHN HADDOCK, Mathematics Department, Memphis State University, Memphis, Tennessee

TOM HAIGH, Mathematics Department, Rose Hulman Institute of Technology, Terre Haute, Indiana

HENRY HERMES, Department of Mathematics, University of Colorado, Boulder, Colorado

CHARLES R. JOHNSON, Institute for Physical Science and Technology, University of Maryland, College Park, Maryland

A. G. KARTSATOS, Department of Mathematics, University of South Florida, Tampa, Florida

J. P. LaSALLE, Lefschetz Center for Dynamical Systems, Division of Applied Mathematics, Brown University, Providence, Rhode Island

ROGER T. LEWIS, Mathematics Department, University of Alabama in Birmingham, Birmingham, Alabama

WADI MAHFOUD, Mathematics Department, Southern Illinois University at Carbondale, Carbondale, Illinois

REZA MALEK-MADANI, Mathematics Department, Southern Illinois University at Carbondale, Carbondale, Illinois

ROGER McCANN, Department of Mathematics and Statistics, Case Western Reserve University, Cleveland, Ohio

ALEXANDER MORGAN, Mathematics Department, University of Miami, Coral Gables, Florida

DEAN A. NEUMANN, Department of Mathematics, Bowling Green State University, Bowling Green, Ohio

BASIL NICHOLS, University of California, Los Alamos Scientific Laboratory, Los Alamos, New Mexico

GERTRUDE OKHUYSEN, Department of Mathematics, Mississippi State University, Mississippi State, Mississippi

MARY E. PARROTT, Mathematics Department, Memphis State University, Memphis, Tennessee

ANDREW T. PLANT, Department of Mathematics, Vanderbilt University, Nashville, Tennessee

GEORGE W. REDDIEN, Mathematics Department, Vanderbilt University, Nashville, Tennessee

G. REISSIG, Institut für Mathematik, Ruhr-Universität Bochum, Federal Republic of Germany

R. REISSIG, Institut für Mathematik, Ruhr-Universität Bochum, Federal Republic of Germany

JAMES A. RENEKE, Department of Mathematical Sciences, Clemson University, Clemson, South Carolina

F. VIRGINIA ROHDE, Department of Mathematics, Mississippi State University, Mississippi State, Mississippi

VILLELLA ROSANNA, Mathematics Institute, University of Oxford, Oxford, England

SHAWKY E. SHAMMA, Faculty of Mathematics and Statistics, The University of West Florida, Pensacola, Florida

A. S. C. SINHA, Departments of Electrical Engineering and Mathematics, Indiana Institute of Technology, Fort Wayne, Indiana

PAUL W. SPIKES, Department of Mathematics, Mississippi State University, Mississippi State, Mississippi

MICHAEL STECHER, Department of Mathematics, Texas A & M University, College Station, Texas

JAMES R. WARD, Department of Mathematics, University of Oklahoma, Norman, Oklahoma

Z. WARSI, Department of Aerophysics and Aerospace Engineering, Mississippi State University, Mississippi State, Mississippi

GLENN WEBB, Department of Mathematics, Vanderbilt University, Nashville, Tennessee

JACKIE WEINBERGER, Department of Mathematics, Memphis State University, Memphis, Tennessee

STABILITY OF DYNAMICAL SYSTEMS

Part I

INVITED ADDRESSES

Chapter 1

ON THE LIMITING EQUATIONS AND INVARIANCE OF TIME-DEPENDENT DIFFERENCE EQUATIONS

ZVI ARTSTEIN*
Division of Applied Mathematics
Brown University
Providence, Rhode Island

INTRODUCTION

In the main series of lectures in the conference Professor LaSalle developed the analogue of Liapunov's theory and the invariance principle for discrete dynamical systems. He showed how the many concepts which were extensively discussed in the context of ordinary differential equations (o.d.e.'s) can be reinterpreted and used in connection with discrete systems, see [3] and [4]. Here we shall focus on one concept, namely the limiting equations. The basic use of the limiting equations was discussed by LaSalle [3]; here we shall examine the situation where the limiting equations are not of the same type as the original equation. This idea was investigated for o.d.e.'s (see [1] and the references therein.) Mathematically our chapter will be self-contained, but we shall hardly discuss

*Present address: Department of Mathematics, The Weizmann Institute of Science, Rehovot, Israel.

motivation or applications. These can be found in [1,3,4] and the references contained there.

We develop the theory for difference relations of the form $x(n+1) \in T(n,x(n))$ where T is a set-valued mapping. This is a generalization of the linear iterates $x(n+1) = g(n,x(n))$. We are interested in the extension not only for the sake of generalization; we shall show how several types of difference equations can be reduced to this general form. Moreover, even if we start with a linear iterate, the limiting equations have in general the form of a difference relation.

By passing to the general form of a difference relation the theory becomes embarrassingly simple. The definitions are straightforward and natural, and the problem of the existence of the limiting equations (the positive precompactness in [1]) disappears. We shall return to this in our concluding remarks.

NOTATIONS AND TERMINOLOGY

Let X be a finite dimensional euclidean space. (This assumption is only for simplicity, all the results and proofs will be valid for a locally compact space if "bounded" is systematically changed to "precompact".) Let J be the set of all integers. Let $T : J \times X \to X$ be a set-valued mapping, i.e., for each $n \in J$ and $x \in X$ a subset $T(n,x)$ (may be empty) of X is given.

A solution of the difference relation

$$x' \in T(n,x) \qquad (*)$$

is a sequence (finite or infinite) $\{x(n) : M < n < N\}$ which satisfies $x(n+1) \in T(n,x(n))$ for $M < n < N-1$. A solution is maximal if either $N = \infty$ (respectively, $M = -\infty$) or $T(N-1,x(N-1))$ is empty (respectively, for no y the relation $x(M+1) \in T(M,y)$ holds). We shall often associate an initial value $x(0) = x_0$ with (*).

We shall always assume the following.

ASSUMPTION H. The mapping T has a closed graph, i.e., for each $n \in J$ the set $\{(x,y) : y \in T(n,x)\}$ is closed in $X \times X$. In particular all the values of T are closed.

If R(.) is a set-valued mapping we write $R^-(.)$ for the set-valued mapping $R^-(x) = \{z : x \in R(z)\}$. If T is as above, we shall denote by T^- the mapping $T^-(n,x) = \{z : x \in T(n-1,z)\}$. Notice that the relation $x' \in T^-(n,x)$ represents a change in the direction of the time variable.

Let $\phi = \{\phi(n)\}$ be a solution of (*). The ω-limit set of ϕ, denoted by $\Omega(\phi)$, consists of all the vectors z such that $z = \lim \phi(k_i)$ where $k_i \to \infty$. In particular if ϕ is defined only for a finite number of positive integers then $\Omega(\phi)$ is empty.

THE INVARIANCE IN THE AUTONOMOUS CASE

Since we generalize in this note the invariance property from the autonomous (= time independent) to the nonautonomous case, we feel an obligation to at least state the former property. Suppose T is autonomous, that is T(n,x) does not depend on n, and we write it as T(x). Let ϕ be a solution of $x' \in Tx$.

A. If $\{\phi(n) : n > 0\}$ is bounded then the ω-limit set $\Omega(\phi)$ is invariant, i.e., for each $z \in \Omega(\phi)$ the sets $T(z) \cap \Omega(\phi)$ and $T^-(z) \cap \Omega(\phi)$ are not empty.

B. If for every bounded set B in X the set $\cup\{T(x) : x \in B\}$ is bounded then $\Omega(\phi)$ is positively invariant, i.e., $T(z) \cap \Omega(\phi)$ is not empty for every $z \in \Omega(\phi)$. If for every bounded B the set $\cup\{T^-(x) : x \in B\}$ is bounded then $\Omega(\phi)$ is negatively invariant, i.e., $T^-(z) \cap \Omega(\phi)$ is not empty for every $z \in \Omega(\phi)$.

It is not difficult to prove the two statements; they follow from Assumption H. At any rate, they are particular cases of their generalizations (Propositions 4 and 5) below. Notice that the positive (and respectively the negative) invariance implies that for $z \in \Omega(\phi)$ the initial value problem $x' \in T(x)$, $x(0) = z$ has a solution defined for all $n > 0$ (respectively $n < 0$).

THE LIMITING EQUATIONS AND INVARIANCE

The following definition, concerning the convergence of the right hand side of (*), will be the basis for introducing the limiting equations below.

DEFINITION 1. The set-valued mapping $T : J \times X \to X$ is the ℓ-limit of the sequence $T^{(m)}$ of set-valued mappings $T^{(m)} : J \times X \to X$ provided $y \in T(n,x)$ if and only if $(x,y) = \lim (x_m,y_m)$, where $y_m \in T^{(m)}(n,x_m)$.

The ℓ in the definition stands for "lower", and the convergence is motivated by the lower limit of a sequence of sets (the graphs of $T^{(m)}$ in our case); see Kuratowski [2; p. 335].

DEFINITION 2. The translate of T by the integer k is the mapping T_k defined by $T_k(n,x) = T(n+k,x)$.

Notice that the initial value problem $x' \in T(n,x)$, $x(k) = x_0$ is equivalent, up to a translate $n \to n+k$ in the domain of the solutions, to the initial value problem $x' \in T_k(n,x)$, $x(0) = x_0$.

DEFINITION 3. The relation $x' \in S(n,x)$ is an ℓ-limiting equation of $x' \in T(n,x)$ if there is a sequence of integers $k_i \to \infty$ such that S is the ℓ-limit of T_{k_i} as $i \to \infty$.

It is easily seen that S satisfies Assumption H.

Remark. Any sequence $k_i \to \infty$ determines a limiting equation, since the ℓ-limit is defined. This is not the situation in o.d.e.'s (see [1,3] and references therein) or if we restrict ourselves to linear iterates (as in [4]). In these cases even the existence of a limiting equation is sometimes in doubt.

We shall now state and prove the two invariance results analogous to A and B.

PROPOSITION 4. Let ϕ be a bounded solution of $x' \in T(n,x)$. Then for each $z \in \Omega(\phi)$ there is an ℓ-limiting equation $x' \in S(n,x)$ such that the initial value problem $x' \in S(n,x)$, $x(0) = z$ has a solution ψ defined on the entire set J and satisfies $\psi(n) \in \Omega(\phi)$ for all $n \in J$.

Proof. The vector z is a limit of a sequence $\phi(k_i)$ with $k_i \to \infty$. We now successively define subsequences as follows. First, $k_{i,0}$ is given by $k_{i,0} = k_i$. Suppose $k_{i,j}$ is given. For each $k_{i,j}$ consider the block $[\phi(k_{i,j}-j-1), \phi(k_{i,j}-j), \ldots, \phi(k_{i,j}+j+1)]$. A subsequence of this sequence of (2j+3) - vectors converges. The indices of this subsequence are denoted $k_{i,j+1}$. Let m run over the diagonal sequence $k_{m,m}$. Let S be the ℓ-limit of T_m. We claim that this S is the desired limiting equation. Indeed, for $n \in J$, if m is large enough then $\psi_m(n) = \phi(m+n)$ is defined and converges, say to $\psi(n)$, as $m \to \infty$. Being the limit of $\phi(m+n)$ as $m \to \infty$ implies that $\psi(n)$ is a member of $\Omega(\phi)$. From the definition of ℓ-limit it then follows that ψ is a solution of $x' \in S(n,x)$, $x(0) = z$.

PROPOSITION 5. If for every bounded B in X the set $\cup\{T(n,x) : n \geq n_0, x \in B\}$ is bounded, then the ω-limit set $\Omega(\phi)$ of any solution ϕ is positively invariant in the sense that for each $z \in \Omega(\phi)$ there is an ℓ-limiting equation $x' \in S(n,x)$ such that the initial value problem $x' \in S(n,x)$, $x(0) = z$ has a solution ψ defined for $n > 0$ and satisfying $\psi(n) \in \Omega(\phi)$ for all $n > 0$. Similarly, the set $\Omega(\phi)$ is negatively invariant if for any bounded B the set $\cup\{T^-(n,x) : n \geq n_0, x \in B\}$ is bounded.

The proof is almost the same as in the preceding proposition. The only change is that in order to establish the positive invariance, we consider blocks of the form $[\phi(k_{i,j}), \ldots, \phi(k_{i,j}+j+1)]$ (and of course $[\phi(k_{i,j}-j-1), \ldots, \phi(k_{i,j})]$ for the negative invariance). The boundedness of these blocks, which guarantees the convergence of the subsequences, is deduced from the boundedness condition in the statement of the proposition.

Remark. We established the invariance with respect to ℓ-limiting equations. Obviously, positive or negative invariance with respect to this class of difference relations implies the respective invariance with respect to any other class of relations provided that the graph of any ℓ-limiting equation is contained in the graph of an element in the corresponding class. Sometimes it may be easier to determine the larger class of equations.

HOW DIFFERENCE RELATIONS ARISE

We have presented the general theory for the difference relation

$$x' \in T(n,x) \tag{*}$$

In this setting the limiting equations have the same form as the original equation and its translates. The situation is different when the original equation has the form of a linear iterate

$$x' = g(n,x) \tag{1}$$

where g is a single-valued map. Equation (1) is of course the particular case of (*) where $T(n,x) = \{g(n,x)\}$. But the limiting equations of (1) do not in general have the form (1). Sufficient conditions for the existence of limiting equations of the form (1) can be found in LaSalle [3].

Another case where difference relations arise is when we deal with a control system

$$x' = g(n,x,c) \tag{2}$$

and the control c takes values in a certain prescribed set C. The admissible trajectories (namely the solutions) are identical with the solutions of the difference relation (*) if we let $T(n,x) = \{g(n,x,c) : c \in C\}$. The limiting equations can (under quite general conditions) be presented in the form (2). We shall not pursue this direction here.

We do want to emphasize here that (*) includes difference equations which are not linear iterates, for example

$$x' = f(n,x,x') \tag{3}$$

A solution of (3) is a function ϕ such that $\phi(n+1) = f(n,\phi(n),\phi(n+1))$, so here we have to "solve" the equation and not just iterate the mapping. (Notice that since (3) is a system, this form includes higher order equations of the form $x(n+1) = f(n,x(n-k),\ldots,x(n),x(n+1))$.) If we define the mapping T in (*) by $T(n,x) = \{y : y = f(n,x,y)\}$ we again obtain a difference relation which is identical, (solution-wise) to (3). The difficulty here is that T is given implicitly. But, basically, we can regard the limiting equations of the difference relation obtained as the limiting equations of the equation (3). It would be better to be able to express these equations in terms of the function f and not in terms of the implicitly given mapping T. To get exactly the ℓ-limiting equations might be too difficult. What can be more easily done is to obtain a class of equations which contain the ℓ-limiting equations in the sense described in the previous remark. We shall not do this here.

Finally we want to point out that we have proved the invariance with respect to a quite large family of limiting equations. It is so rich that we do not have the problem of the existence of a limiting equation. If we want to establish the invariance with respect to a certain subclass E, it will be enough to show that for every sequence of integers $k_i \to \infty$ a subsequence m_i exists such that the ℓ-limiting equation determined by that subsequence belongs to the subclass E. This property can be regarded as positive precompactness with respect to E. The conditions for positive precompactness in [3] give the result for the equation (1) and the class E of equations of the same form.

REFERENCES

1. Z. Artstein, Limiting equations and stability of non-autonomous ordinary differential equations, in The Stability of Dynamical Systems, (Appendix A), CBMS Regional Conference Series in Applied Mathematics, Vol. 25, SIAM, Philadelphia, 1976.

2. K. Kuratowski, Topology I, Academic, New York, 1966.

3. J. P. LaSalle, The Stability of Dynamical Systems, CBMS Regional Conference Series in Applied Mathematics, Vol. 25, SIAM, Philadelphia, 1976.

4. J. P. LaSalle, Stability of difference equations; in A Study in Ordinary Differential Equations, (edited by J. K. Hale), Studies in Mathematics Series, Mathematical Association of America, to appear.

Chapter 2

LIAPUNOV FUNCTIONS AND BOUNDEDNESS OF SOLUTIONS

T. A. BURTON
Department of Mathematics
Southern Illinois University
Carbondale, Illinois

INTRODUCTION

We consider a system of ordinary differential equations

$$X' = F(t,X) \qquad (' = d/dt) \tag{1}$$

where $F : [0,\infty) \times R^n \to R^n$ with F continuous. Assume that there is a function $V : [0,\infty) \times R^n \to [0,\infty)$ with continuous first partial derivatives and with $V'(t,X) \leq 0$ along solutions of (1).

If $V(t,X) \to \infty$ as $|X| \to \infty$ uniformly for $0 \leq t < \infty$, then V is said to be radially unbounded. In this case, it is well known that all solutions of (1) are bounded in the future. However, often V is not radially unbounded and the problem here is to then see what can be salvaged.

Radial unboundedness is merely a convenience for proving boundedness results. A considerably more fundamental property showing boundedness is the angle between the vectors F and grad V, at least when V is autonomous. In that case, $V' = \text{grad } V \cdot F = |\text{grad } V||F|\cos\theta$, and if there are unbounded sets satisfying $V(x)$ = constant, along which $\cos \theta \leq -\varepsilon$ for some $\varepsilon > 0$, no solution can become unbounded along that set.

We generalize this idea and obtain a boundedness result which we then apply to the problem of Lurie.

The reader is referred to Hahn [7] for a general treatment of Liapunov's direct method. Complete details of the results in this chapter will appear elsewhere in [3].

BOUNDEDNESS

It is assumed that there are k disjoint unbounded sets on which V may be bounded and through which a solution could escape. Each such set may be distinguished by a "level" surface $V(t,X) = L_i$, $i = 1,\ldots,k$. We first augment V by adding a function μ so that $V(t,X) + \mu(X)$ is radially unbounded. This is the affect of Definition 1. The function μ is not continuous except on certain open sets and so $V + \mu$ is not a Liapunov function. However, one can show that if $\text{grad}\,\mu \cdot F \leq 0$ when $\text{grad}\,\mu$ is defined, then the solutions of (1) are bounded. That is the content of Theorem 0. If $\text{grad}\,\mu \cdot F \leq 0$ fails, then we ask essentially that $[\text{grad}\,V \cdot F]/|\text{grad}\,V||F|$ not decrease too rapidly as $|X|$ increases in certain subsets of those sets in which V is bounded. That is, we ask that $\cos \theta$ not approach zero too rapidly. This condition is formalized as Definition 2.

Below, $\overline{R}$ denotes the closure of R and R^c denotes the complement.

<u>DEFINITION 1.</u> A function $V : [0,\infty) \times R^n \to [0,\infty)$ is augmented by μ if there is a function $\mu : R^n \to [0,\infty)$ such that $V(t,X) + \mu(X)$ is radially unbounded and if the following two conditions hold.

a. There are disjoint open sets $R_1,\ldots,R_k$ in R^n and continuous functions $\mu_1,\ldots,\mu_k$ with $\mu_i : \overline{R}_i \to [0,\infty)$. Each μ_i has continuous first partial derivatives in R_i, and

$$\mu(X) = \begin{cases} \mu_i(X) \text{ if } X \in R_i \text{ for some } i \\ 0 \text{ if } X \in (\cup R_i)^c \end{cases}$$

b. There are positive constants $L_1,\ldots,L_k$ such that for each i, if $0 < L_i^* < L_i$, then there exists $D > 0$ such that if

$$X \in R_i \text{ and } V(t,X) \le L_i^*, \text{ then } \mu_i(X) \le D \qquad (2)$$

Note that since $V(t,X) + \mu(X)$ is radially unbounded, for each $L > 0$, there exists $H > 0$ such that

$$\text{if } V(t,X) \le L \text{ and } |X| \ge H, \text{ then } X \in R_i \text{ for some } i \qquad (3)$$

The following result is well-known, at least for second order systems, and has been used in many examples. The present formulation, however, is both new and useful. Note that part b of Definition 1 is not required here.

THEOREM 0. Suppose that V is augmented by μ according to Definition 1. If for each $X \in \cup R_i$ we have $\text{grad}\, \mu(X)\cdot F(t,X) \le 0$ for all $t \ge 0$, then all solutions of (1) are bounded.

The proof will appear in [3].

DEFINITION 2. A function V is an augmented Liapunov function if V is augmented by μ according to Definition 1 and if for each i and for each $L > L_i$, there exists a positive constant J and continuous functions $g : (0, L - L_i] \to (0,\infty)$ and $h : [J,\infty) \to [0,\infty)$ with

$$\int_{0^+}^{L-L_i} [ds/g(s)] < \infty \quad \text{and} \quad \int_J^\infty h(s)ds = \infty$$

while $\mu_i(X) \ge J$ and $L \ge V(t,X) > L_i$ imply

$$V'(t,X) \le -g(V(t,X) - L_i)h(\mu_i(X))|\text{grad}\, \mu_i(X)\cdot F(t,X)|$$

THEOREM 1. Suppose that V is an augmented Liapunov function according to Definition 2 and that for each i, if $\mu_i(X) \ge J$ and $V(t,X) = L_i$, then $V'(t,X) < 0$. Then all solutions of (1) are bounded.

The proof will appear in [3].

Since this result is fairly general and, as a consequence, is quite complicated, we will state and prove a simplified version.

THEOREM 2. Suppose that the following conditions hold.

I. There exists $L > 0$ such that $0 < L^* < L$ and $V(t,X) \leq L^*$ imply $|X|$ bounded.

II. There exists $J > 0$ such that $V(t,X) = L$ and $|X| \geq J$ imply $V'(t,X) < 0$.

III. For each $\varepsilon > 0$, there exist $g : (0,\varepsilon) \to (0,\infty)$ and $h : [J,\infty) \to [0,\infty)$ such that $L + \varepsilon > V(t,X) > L$ and $|X| \geq J$ imply:

i. $V'(t,X) \leq -g(V(t,X) - L)h(|X|)|F(t,X)|$, and

ii. $\int_{0^+}^{\varepsilon} [ds/g(s)] < \infty$ and $\int_J^{\infty} h(s)ds = \infty$.

Then all solutions of (1) are bounded.

Proof. If the theorem is false, then there is an unbounded solution $X(t)$ on a right maximal interval $[t_0,T)$. We can argue from II and then I that $V(t, X(t)) > L$. We choose ε by setting $V(t_0, X(t_0)) = L + \varepsilon$ and obtain g and h from III.

Either $|X(t)| \geq J$ on some interval $[t_1,T)$ or there are sequences $\{t_n\}$ and $\{T_n\}$ with $t_n < T_n < t_{n+1}$, $|X(t_n)| = J$, $|X(t)| \geq J$ on $[t_n, T_n]$, and $|X(T_n)| \to \infty$ as $n \to \infty$.

In the first case, we separate variables in IIIi and integrate from t_1 to $t > t_1$, to obtain

$$\begin{aligned}\int_{t_1}^{t} &[V'(s, X(s))/g(V(s, X(s)) - L)]ds \\ &\leq -\int_{t_1}^{t} h(|X(s)|)|X(s)|ds \\ &\leq -\int_{t_1}^{t} h(|X(s)|)||X(s)|'|ds\end{aligned}$$

Changing variables on both sides of the inequality yields

$$\int_{r(t_1)}^{r(t)} [ds/g(s)] \leq -\left| \int_{|X(t_1)|}^{|X(t)|} h(u)du \right|$$

where $r(t) = V(t, X(t)) - L$. This yields $X(t)$ bounded. A similar argument on the intervals $[t_n, T_n]$ completes the proof.

Remark 1. The present author [1, 2] attempted to show boundedness by requiring that V' have certain strong properties. It was subsequently shown by Erhart [5; Th. 2.1] that the property on V' generally implied V radially unbounded. The example in the next section shows that our conditions hold without V being radially unbounded.

Remark 2. If F is bounded for X bounded, then results of LaSalle [8] and Haddock [6] yield interesting additional information about solutions. If some component of F is bounded for X bounded, then the results in [4] yield similar additional information.

A LURIE PROBLEM

We consider the system

$$x' = Ax + b\phi(\sigma)$$
$$\sigma' = c^T x - r\phi(\sigma)$$

in which $\phi : R \to R$, $\sigma\phi(\sigma) > 0$ if $\sigma \neq 0$, ϕ is continuous, A is an $n \times n$ matrix all of whose characteristic roots have negative real parts, c and b are constant n-vectors. The reader is referred to Lefschetz [10; pp. 20-21] for a general discussion of the problem.

One uses

$$V(x,\sigma) = x^T Bx + \phi(\sigma)$$

where $\phi(\sigma) = \int_0^\sigma \phi(s)ds$ and B is positive definite and symmetric, obtaining

$$V' = -x^T Dx + \phi(\sigma)[2b^T B + c^T]x - r\phi^2(\sigma)$$

which is negative definite in $(x,\phi(\sigma))$ when $D = -(A^T B + BA)$ and

$$r > (Bb + c/2)^T D^{-1}(Bb + c/2) \tag{4}$$

We note that if $\phi(\pm\infty) \neq \infty$, then V is not radially unbounded. For brevity, take $\phi(\infty) = \phi(-\infty)$ so that the choice of $L_1 = L_2 = \phi(\infty)$ will satisfy (2) and (3) when $R_1 = \{(x,\sigma) : \sigma > 0\}$, $R_2 = \{(x,\sigma) : \sigma < 0\}$, $\mu_1(x,\sigma) = \sigma$, and $\mu_2(x,\sigma) = -\sigma$.

If (4) holds, then V' negative definite in $(x, \phi(\sigma))$ implies that there exists $m > 0$ with $V' \leq -m(x^T x + \phi^2(\sigma))$. Now $V(x,\sigma) - \phi(\infty) = x^T Bx + \phi(\sigma) - \phi(\infty) \leq x^T Bx \leq Qx^T x$ for some $Q > 0$. Also,

$$|\sigma'| = |c^T x - r\phi(\sigma)| \leq P[x^T x + \phi^2(\sigma)]^{1/2}$$

for some $P > 0$.

Thus, we pick $h(s) = m/P\sqrt{Q}$ and g the square root function since we have

$$V' \leq -(m/P\sqrt{Q})[V(x,\sigma) - \phi(\infty)]^{1/2}|\mu_i'|$$

Theorem 1 then yields all solutions bounded.

We note that there is also a largely algebraic proof of this result given by LaSalle [9].

REFERENCES

1. T. A. Burton, An extension of Liapunov's direct method, J. Math. Anal. Appl. 28 (1969), 545-552.
2. T. A. Burton, Correction to An extension of Liapunov's direct method, J. Math. Anal. Appl. 32 (1970), 688-691.
3. T. A. Burton, Liapunov functions and boundedness, to appear in J. Math. Anal. Appl.
4. T. A. Burton and J. W. Hooker, On solutions of differential equations tending to zero, J. reine angew. Math. 267 (1974), 151-165.
5. J. V. Erhart, Lyapunov theory and perturbations of differential equations, SIAM J. Math. Anal. 4 (1973), 417-432.
6. J. R. Haddock, On Liapunov functions for nonautonomous systems, J. Math. Anal. Appl. 47 (1974), 599-603.
7. W. Hahn, Theory and Applications of Liapunov's Direct Method, Prentice-Hall, Englewood Cliffs, 1963.
8. J. P. LaSalle, Stability theory for ordinary differential equations, J. Differential Equations 4 (1968), 57-65.
9. J. P. LaSalle, Complete stability of a nonlinear control system, Proc. Nat. Acad. Sci. U.S.A. 48 (1962), 600-603.
10. S. Lefschetz, Stability of nonlinear control systems, Math in Sci. and Engineering, Vol. 13, Academic, New York, 1965.
11. G. Sansone and R. Conti, Non-linear Differential Equations, Pergamonn, New York, 1964.
12. R. Reissig, G. Sansone, and R. Conti, Qualitative Theorie Nichtlinear Differentialgleichungen, Edizioni Cremonese, Rome, 1963.

Chapter 3

RECENT RESULTS ON OSCILLATION OF SOLUTIONS OF FORCED AND PERTURBED NONLINEAR DIFFERENTIAL EQUATIONS OF EVEN ORDER

ATHANASSIOS G. KARTSATOS

Department of Mathematics
University of South Florida
Tampa, Florida

INTRODUCTION

The main purpose of this paper is to present some modern results pertaining to the oscillation of solutions of forced and perturbed nonlinear equations. We are mostly concerned here with integral criteria guaranteeing oscillation. The interest of mathematicians in this direction was triggered by the now classical paper of Atkinson [2] in 1955 concerning the important case of an Emden-Fowler-like equation of second order.

For two reasons we almost always consider, with few exceptions, only even order equations. First, we avoid statements concerning n even and n odd separately; and second, those familiar with the theory will be able to interpret the results for n odd without any difficulty. Many of the results presented here were originally formulated for more complicated and/or functional equations. We gave a simplified version in order to make the exposition easier. An extensive bibliography on the subject is also given at the end of the paper. Some "linear" papers affecting the nonlinear case are included therein; some others have been excluded and might be found in the bibliography of Wong's second order survey article [342]. The paper is divided into six sections. The first is devoted to the preliminaries. In the second we consider the homogeneous case whose oscillatory behavior definitely affects the behavior of the forced and the perturbed case. The third and fourth sections are concerned with the forced and the perturbed case respectively. In the fifth section some new results are given, and the last section is devoted to the formulation of several problems which, according to the opinion of this author, are still open.

The author wishes to express his thanks to Professor J. R. Graef of the Mississippi State University for the opportunity of presenting an invited address, and his, as well as Professor Spikes', hospitality. The author also wishes to apologize to those of his colleagues whom he did not mention either in the text, or in the bibliography, due to his ignorance concerning their work.

PRELIMINARIES

In what follows we let $R = (-\infty, \infty)$, $R_+ = [0,\infty)$, $R_+^o = R_+ - \{0\}$, $R_- = (-\infty,0]$, $R_-^o = R_- - \{0\}$. The letter T will denote a fixed nonnegative number, and $R_T = [T,\infty)$. Without further mention all the functions considered will be assumed continuous on their respective domains. Unless otherwise stated, the letter n will always denote an even integer ≥ 2. Now consider the equation

$$x^{(n)} + H(t,x,x',\dots,x^{(n-1)}) = 0 \tag{1.1}$$

with n arbitrary and $H : R_T \times R^n \to R$. By a solution of (1.1) we mean any real function $x(t)$ which is n times continuously differentiable on an interval $[t_x,\infty)$ ($t_x \geq T$) and satisfies (1.1) on this interval. The number t_x depends on the particular solution $x(t)$. A function $f : [a,\infty) \to R$ ($a \geq T$) is said to be oscillatory if it has an unbounded set of zeros on $[a,\infty)$. Equation (1.1) is said to be oscillatory if all of its solutions are oscillatory. A function f as above is said to be bounded if there exists a positive constant k such that $|f(t)| \le k$, $t \ge a$. Equation (1.1) is said to be B-oscillatory if all of its bounded solutions are oscillatory. Only non-eventually trivial solutions are considered here.

Now we consider the following three types of equations:

$$x^{(n)} + H(t,x) = 0 \tag{I}$$

$$x^{(n)} + H(t,x) = Q_1(t) \tag{II}$$

$$x^{(n)} + H(t,x) = Q_2(t,x) \tag{III}$$

The functions H, Q_1, Q_2 will occasionally satisfy one of the following hypotheses:

(H) $H : R_T \times R \to R$, $uH(t,u) > 0$ for every $u \neq 0$.

(Q_1) $Q_1 : R_T \to R$ and there exists a function $P : R_T \to R$, which is n times continuously differentiable on R_T with $P^{(n)} = Q_1(t)$, $t \in R_T$.

(Q_2) $Q_2 : R_T \times R \to R$ and there exists a continuous function $Q_o : R_T \times R_+ \to R_+$ such that $|Q_2(t,u)| \le Q_o(t,|u|)$ for every $(t,u) \in R_T \times R$.

Let us now show that a nonlinear equation of order two may have oscillatory and nonoscillatory solutions. This behavior has been exhibited by Moore and Nehari [227]. In fact, it was shown there that the equation

$$x'' + p(t)g(x) = 0 \tag{1.2}$$

with $p(t) \equiv (1/4)t^{-(m+2)}$, $g(u) \equiv u^{2m+1}$, $t \ge 1$, m a natural number, has the following three types of solutions: (i) x(t) is nonoscillatory and tends to a finite limit as $t \to \infty$; (ii) x(t) is nonoscillatory, monotone, and such that the function $x(t) - t^{1/2}$, $t \ge 1$ is oscillatory; (iii) x(t) is oscillatory. This example implies immediately that Sturm's comparison theorem does not hold in its full generality even for second order equations. It is also proper to note here that an equation of the form (1.1) might have nonextendable solutions for which our definition of oscillation does not apply. In fact, in the interesting and extensively studied case (1.2), Burton and Grimmer [33] proved the following:

THEOREM 1.1. (Burton and Grimmer [33]). Consider (1.2) with $p : R_T \to R$, $g : R \to R$ and $ug(u) > 0$ for every $u \neq 0$. Let $p(t_1) < 0$ for some $t_1 \geq T$. Then if anyone of the following conditions holds, (1.2) has solutions which are not continuable to $+\infty$:

$$\text{(a)} \quad \int_0^{\infty} [1 + G(s)]^{-1/2} ds < +\infty,$$

$$\text{(b)} \quad \int_0^{-\infty} [1 + G(s)]^{-1/2} ds > -\infty,$$

where $G(u) = \int_0^u g(v)\,dv$.

The sufficiency part of this result has been shown to hold in the arbitrary nth order case by Mahfoud [218] and for large classes of functions H as in (1.1). In an unpublished result, Prof. Burton has shown the necessity part to be also true in third order equations. In this connection, the reader is also referred to Kiguradge [161], who, among other results, establishes conditions under which a certain particular form of (1.2) has some extendable oscillatory solutions despite the oscillatory character of $p(t)$. In this same paper, Kiguradge shows the sufficiency part of the above theorem for a special case again of (1.2). Nevertheless, under (H), all nonoscillatory solutions of (I) are extendable to $+\infty$, and a reference to the proof of this fact is Foster's paper [73]. The uniqueness problem will not concern us here because it hardly ever enters into the research for oscillation criteria in the present spirit.

THE HOMOGENEOUS CASE

In this section we shall start with the pioneering result of Atkinson [2] and continue with some of the highlights of its extensions.

THEOREM 2.1. (Atkinson [2]). Consider the equation

$$x'' + p(t)x^{2m+1} = 0 \tag{2.1}$$

where $p : R_T \to R_+^o$ and m is a positive integer. Then the condition

$$\int_T^\infty tp(t)\,dt = +\infty \tag{2.2}$$

is necessary and sufficient for (2.1) to be oscillatory.

Atkinson [2] also gave a result whose conclusion ensures the nonoscillation of all solutions of (2.1):

THEOREM 2.2. (Atkinson [2]). Let p(t) be positive and continuously differentiable on R_T with $p'(t) \leq 0$ there. Let m be a nonnegative integer. If

$$\int_T^\infty t^{2m+1}p(t)\,dt < +\infty$$

then (2.1) has no oscillatory solutions.

The proof of the above theorem is based on the fact that all solutions of (2.1) have bounded derivatives under these hypotheses. As far as the author knows, there is no analogue to this theorem covering the nth order case (cf. Problem I). The present author [138] gave a result which ensures the nonoscillation of all bounded solutions of nth order superlinear equations:

THEOREM 2.3. (Kartsatos [138]). Consider the equation

$$x^{(n)} + H(t,x,x',\ldots,x^{(n-1)}) = 0 \tag{2.3}$$

where $t \in R_T$, $H : R_T \times R^n \to R$, and such that

$$|H(t,u(t),\ldots,u^{(n-1)}(t))| \leq H(t;u)|u(t)|$$

$$\int_T^\infty t^{n-1}H(t;u)\,dt < +\infty$$

for every bounded function u which is defined and n times continuously differentiable on R_T. Here H maps R_T into R_+ and depends on u. Then every bounded oscillatory solution of (2.2) is identically equal to zero for all large t.

Ličko and Švec [203] gave the following result.

THEOREM 2.4. (Ličko and Švec [203]). Consider the equation (I) with $H(t,u) \equiv P(t)u^{\alpha}$, where $p(t)$ is positive on R_T and α is the quotient of two odd positive integers. Let n be even and $0 < \alpha < 1$. Then the condition

$$\int_T^{\infty} t^{\alpha(n-1)} p(t)dt = +\infty \tag{2.4}$$

is necessary and sufficient for the oscillation of Equation (I). Let n be odd, $0 < \alpha < 1$. Then (2.4) is necessary and sufficient for all solutions of (I) to oscillate or tend monotonically to zero. In the case $\alpha > 1$ both these conclusions hold if (2.4) is replaced by

$$\int_T^{\infty} t^{n-1} p(t)dt = +\infty \tag{2.5}$$

The sufficiency part of the above criterion for n even and $\alpha > 1$ was actually given for the first time by Kiguradge [157; Theorem 5] in 1962. The case n even and $0 < \alpha < 1$ in the above theorem extends a result of Belohorec [6] who considered second order equations. In [132], the author considered equations of the form (2.3) with H jointly homogeneous with respect to its last n variables, and established the following.

THEOREM 2.5. (Kartsatos [132]). Consider the equation (2.3) and assume that $H(t,u_1,\ldots,u_n) \equiv p(t)g(u_1,u_2,\ldots,u_n)$, where $p : R_T \to R_T^{\,o}$, $g : R^n \to R$, $u_1 g(u_1,u_2,\ldots,u_n) > 0$ for $u_1 \neq 0$ and such that for some positive constant K and for every $(u_1,u_2,\ldots,u_n) \in R^n$ and every $\lambda \geq K$, $g(-u_1,-u_2,\ldots,-u_n) = -g(u_1,u_2,\ldots,u_n)$ and $g(\lambda, \lambda u_2,\ldots,\lambda u_n) = \lambda^{\alpha} g(1,u_2,\ldots,u_n)$, where $\alpha = q/r$ with q and r odd positive integers. Then, under anyone of the following conditions, (a) for n even, (2.3) is oscillatory, (b) for n odd, every solution of (2.3) oscillates or tends monotonically to zero as $t \to \infty$:

$$\text{(i)} \quad \alpha < 1, \quad \int_T^{\infty} t^{\alpha(n-1)} p(t)\,dt = +\infty$$

$$\text{(ii)} \quad \alpha = 1, \quad \int_T^{\infty} t^{n-1-\varepsilon} p(t)\,dt = +\infty, \text{ for some } \varepsilon > 0$$

$$\text{(iii)} \quad \alpha > 1, \quad \int_T^{\infty} t^{n-1} p(t)\,dt = +\infty$$

The case (ii) above extends a linear theorem of Mikusinski [224]. Mikusinski claims that the conclusion (b) above holds in the case (ii) without ε in the integral condition. This is false, however, and a counterexample was given by the author in [134]:

$$x^{(n)} + p(t)x = 0, \quad t \geq 1, \quad n = \text{odd}, \tag{2.6}$$

with $p(t) \equiv m(m - 1)(m - 2)\cdots(m - (n-2))(n - 1 - m)t^{-n}$, where $n - 2 < m < n - 1$. A family of solutions of (2.6) is given by $x_C(t) = Ct^m$ for every $C \neq 0$, while (2.6) satisfies the assumptions of Mikusinski's theorem. For an earlier example the reader is referred to Ananeva and Balaganskii [1].

Equations like the ones considered in the above theorem with homogeneous g are rather interesting. Their fundamental properties in the second order case were studied by Bihari in [18-20]. From the proof of the above theorem, it is easy to see that almost all the known criteria for oscillation of equations of the type (I) with $H(t,u) \equiv p(t)u^{\alpha}$ (α as above) imply corresponding ones for homogeneous functions H provided that $p(t) \geq 0$. With this in mind, it becomes evident that some of the results of Gustafson [97] are immediate consequences of the proof of the above theorem. Equations (I) with homogeneous-like sublinear H have also been considered by Kiguradge in [160]. Several extensions of Theorem 2.5 were given by Onose in [238, 239, 243, and 244]. The reader is also referred to the paper of Staikos and Sficas [291] for further extensions to functional equations. It should also be mentioned here that Theorem 2 in Ryder and Wend [258] can also be proved as the above theorem, and their case is covered by the remarks of the author on page 602 of [132].

It is convenient now to state a theorem which was established by the author in [140]. This result extends the famous Sturm's comparison theorem (in a certain direction), and, in addition to being important in its own right, appears to be a valuable tool in oscillation arguments. This result was inspired by a certain theorem of Atkinson [3] which pertains to the existence of positive solutions of homogeneous second order equations.

THEOREM 2.6. (Kartsatos [140]). Consider the two equations

$$x^{(n)} + H_i(t,x) = Q_1(t), \; i = 1, 2, \tag*{(2.7)$_i$}$$

where each H_i satisfies (H) (with H replaced by H_i) and is increasing with respect to the second variable. Let Q_1 satisfy (Q_1) with P(t) oscillatory and such that $\lim_{t\to\infty} P(t) = 0$. Moreover, let

$$H_1(t,u) \leq H_2(t,u) \text{ if } t \in R_T \text{ and } u > 0 \tag{2.8}$$

$$H_1(t,u) \geq H_2(t,u) \text{ if } t \in R_T \text{ and } u < 0 \tag{2.9}$$

and the equation $(2.7)_1$ be oscillatory. Then this is also the case for the equation $(2.7)_2$.

Actually, the assumptions (2.8), (2.9) of this theorem can be weakened to the following:

$$H_1^*(t,u) \leq H_2(t,u) \text{ if } t \in R_T \text{ and } u > 0 \tag{2.10}$$

$$H_2^*(t,u) \geq H_2(t,u) \text{ if } t \in R_T \text{ and } u < 0 \tag{2.11}$$

where $H_1^* : R_T \times R_+^o \to R_+^o$ and $H_2^* : R_T \times R_-^o \to R_-^o$ and increasing in u, and such that the equation

$$x^{(n)} + H_1^*(t,x) = Q_1(t) \tag{2.12}$$

has no positive solution and the equation

$$x^{(n)} + H_2^*(t,x) = Q_1(t) \tag{2.13}$$

has no negative solution. The conclusion of the theorem remains the same. Of course $(2.7)_1$ does not play any role now.

Having this improvement in mind, one can easily deduce the main result of Jones ($H(t,u) \equiv \sum_{i=1}^{k} p_i(t)u^i$) in [117] from the main theorem of Atkinson [2], the main theorem of Macki and Wong [217] from the oscillation theorem of Pinter [256], and Theorem 1 of Ryder and Wend [258] from Theorem 2 of Kartsatos [133] (cf. note added in proof of [133]). Similarly, a host of other oscillation criteria can now be obtained by this comparison theorem, but they are too many to enumerate. One of the first results on the oscillation of (I) with nonhomogeneous H was given by Kiguradge in [157] and concerns the superlinear case.

THEOREM 2.7. (Kiguradge [157]). Consider (I) with $H(t,u) = F(t,u^2)u$ where $F(t,v)$ is defined on $R_T \times R_+$ and is non-negative and increasing in v. Moreover, assume that there exists a function $g : R_+^{\,o} \to R_+^{\,o}$ such that $g'(u) \geq 0$, and $F(t,u^2) \geq F(t,c^2)g(u)$ for every $u \in R_+^{\,o}$ and every $c > 0$ with $u > c$. Furthermore, suppose that for every $\varepsilon > 0$,

$$\int_{\varepsilon}^{\infty}[sg(s)]^{-1}ds < +\infty, \quad \int_{T}^{\infty}F(s,c^2)s^{n-1}ds = +\infty$$

Then, for n even, (I) is oscillatory, and, for n odd, every solution of (I) oscillates or tends monotonically to zero as $t \to \infty$.

The author showed in [133] that the conclusion of the above theorem remains valid if $H(t,u) \equiv p(t)g(u)$ with $p(t) > 0$, $ug(u) > 0$ for $u \neq 0$, $g'(u) \geq 0$ for $|u| \geq K > 0$ and

$$\int_{T}^{\infty}t^{n-1}p(t)dt = +\infty, \quad \int_{\varepsilon}^{\infty}[g(s)]^{-1}ds < +\infty, \quad \int_{-\infty}^{-\varepsilon}[g(s)]^{-1}ds < +\infty \tag{2.14}$$

for some $\varepsilon > 0$. The assumption on the differentiability of g(u) was replaced later by a monotonicity assumption and the use of Riemann-Stieltjes integrals by the author in [136].

The integral condition on $p(t)$ above is also necessary in the case of n even, and this was shown by the author in [134], and, independently, by Onose in [240]. Onose considered also the case n even, and his result is the following:

<u>THEOREM 2.8.</u> (Onose [240]). Suppose that n is even, $p : R_T \to R_+$ is bounded, and for the function $g : R^n \to R$ we have $u_1 g(u_1, u_2, \ldots, u_n) > 0$ for every $u_1 \neq 0$. Then the first of (2.14) is necessary and sufficient for the equation

$$x^{(n)} + p(t)g(x, x', \ldots, x^{(n-1)}) = 0$$

to oscillate.

Onose used a successive approximation technique by which he obtained a positive solution to an integral equation associated with (I), under the assumption that the first integral in (2.14) converges. This technique is quite old of course, and an interesting but rather complicated application of it for nonlinear nth order equations was given by Villari [333]. In [134], the author, inspired by the results of Švec [302, 303], used Schauder's fixed point theorem. In the functional case, Schauder's fixed point theorem was applied by Staikos and Sficas [292]. Naturally, mention should be made of the various asymptotic results that Kiguradge obtained, for example, in [159, 160]. In this connection, Pui-Kei Wong's papers [346, 347] are also of importance. In addition to [157], sublinear analogues of Theorem 2.7 above can be found, for example, in the paper [181] by Kusano and Onose. Their result follows however from the corresponding result of Ličko and Švec [203; Theorem 2.6] by use of the comparison result, Theorem 2.6, above. Three years ago, Staikos and Sficas [291] established a so-called "fundamental principle". This principle is the statement, in a compact form, of a method that had been used previously by many authors. In the case of (I) with $H \equiv p(t)g(u)$ and (1.1), it simply asserts that information about (I) can be used to obtain information about (1.1) if we write it as

$$x^{(n)} + [H(t, x, x', \ldots, x^{(n-1)})/g(x)]g(x) = 0$$

provided that $x(t)$ belongs to a certain function class. This idea has been exploited, for example, by Wong [342], Kartsatos [132], and Onose [240], and it works mainly with integral criteria on $p(t)$ which carry over to H/g. We now take a brief look at the notion of strong continuity. This notion was introduced by J. S. W. Wong [342] for second order equations and, independently, by the author [134] for nth order equations. Although the two concepts are different, they do overlap in special cases.

We state here the notion introduced by the author in [134] because it is simpler and we refer the reader to the paper [243] of Onose, who extended Wong's concept to functions of more than two variables, and applied it to nth order equations. Actually, we state here a slightly more general concept than the one in [134].

DEFINITION. (Kartsatos [134]). Assume that $H : R_T \times R^n \to R$, $u_1 H(t,u_1,u_2,\ldots,u_n) > 0$ for every $u_1 \neq 0$, H is increasing in $u_n \in R_+$ and decreasing in $u_n \in R_-$ and the following condition holds:

> Condition S. For every $\varepsilon > 0$ and $\alpha > \varepsilon$ there exists $T_{\alpha,\varepsilon} \geq T$ and a function $P_1(\cdot,\alpha,\varepsilon) : [T_{\alpha,\varepsilon},\infty) \to R_+$ $(P_2(\cdot,\alpha,\varepsilon) : [T_{\alpha,\varepsilon},\infty) \to R_-)$ such that $H(t,x_1,\bar{x},0) \geq P_1(t,\alpha,\varepsilon)$ $(H(t,x_1,\bar{x},0) \leq P_2(t,\alpha,\varepsilon))$ for any $\bar{x} = (x_2,x_3,\ldots,x_{n-1}) \in R^{n-2}$ with $|x_i| < \varepsilon$, $i = 2,3,\ldots,n-1$, any $x_1 \in R$ with $x_1 > \alpha-\varepsilon$ $(x_1 < -\alpha+\varepsilon)$ and any $t \geq T_{\alpha,\varepsilon}$.

Then H is said to be a strongly continuous function.

The following theorem of the author is a new result and a better version of a theorem in [134]; it ensures the oscillation of all the bounded solutions of Equation (1.1).

THEOREM 2.9. (Kartsatos [134]). Let the function H in (1.1) be strongly continuous and such that

$$\int_{T_{\alpha,\varepsilon}}^{\infty} t^{n-1} P_1(t,\alpha,\varepsilon)\,dt = +\infty \quad \text{and} \quad \int_{T_{\alpha,\varepsilon}}^{\infty} t^{n-1} P_2(t,\alpha,\varepsilon)\,dt = -\infty$$

Then (1.1) is B-oscillatory.

The proof of this theorem is trivial if one uses the comparison Theorem 2.6. In fact, let $x(t)$, $t \geq T_1 \geq T$, be a positive solution of (1.1). Then, if $x(t)$ is also bounded, all of its derivatives are of constant sign (up to the order n), $\lim_{t\to\infty} x^{(j)}(t) = 0$, $j = 1,2,\ldots,n-2$, and $x^{(n-1)}(t) \geq 0$ for all large t. If $\lim_{t\to\infty} x(t) = L > 0$, then taking $\varepsilon > 0$ with $\varepsilon < L$, there exists $\bar{t} = T_{L,\varepsilon} \geq T_1$ such that $L-\varepsilon < x(t) < L$, $|x^{(j)}(t)| < \varepsilon$, $j = 1,2,\ldots,n-2$ and $x^{(n-1)}(t) \leq 0$ for every $t \geq \bar{t}$. Now from (1.1) we get

$$x^{(n)}(t) + [H(t,x(t),\ldots,x^{(n-1)}(t))/x(t)]x(t) = 0 \tag{2.15}$$

where $[H(t,x(t),\ldots,x^{(n-1)}(t))/x(t)] \geq (1/L)P_1(t,L,\varepsilon)$ for all $t \geq \bar{t}$. The contradiction can be obtained now from Theorem 2.6 and the fact that the equation

$$x^{(n)} + (1/L)P_1(t,L,\varepsilon)x(t) = 0 \tag{2.16}$$

for $t \geq \bar{t}$, cannot have bounded positive solutions (c.f. Kartsatos [133]). A similar proof holds if $x(t)$ is negative for $t \geq T_1$.

The reader is referred again to the paper of Onose [243], where necessary conditions are also obtained for strongly continuous functions. Obviously, the above definition of strong continuity, as well as the one given by Onose in [243], can be applied only in arguments concerning solutions with bounded derivatives. More general definitions can be given to guarantee the oscillation of all solutions of (1.1), but, again the main role would be played by the comparison Theorem 2.6. It might be helpful to know that the monotonicity assumption of H above with respect to its last variable can be replaced by a condition like the one imposed on $x_2,x_3,\ldots,x_{n-1}$. With this in mind, the above definition of strong continuity applies to the function $H \equiv p(t)g(u_1,u_2,\ldots,u_n)$ with $p(t) > 0$, $x_1 g(u_1,\ldots,u_n) > 0$ for $x_1 \neq 0$. This is also noted by Onose in [243] concerning the concept introduced there. Taking this remark into consideration, let us assume that H is strongly continuous according to the above definition, but without any monotonicity assumption on the last variable. Moreover, $H(t,x_1,\bar{x},0)$ is replaced now by $H(t,x_1,x_2,\ldots,x_n)$. Then we have the following result.

THEOREM 2.10. (Kartsatos [134], Onose [243]). Let H be strongly continuous from the left ($H \geq P_1$) or from the right ($H \leq P_2$). Then Equation (1.1) has a nonoscillatory solution if, for some α and ε

$$\int_{T_{\alpha,\varepsilon}}^{\infty} t^{n-1}P_1(t,\alpha,\varepsilon)\,dt < +\infty, \quad \int_{T_{\alpha,\varepsilon}}^{\infty} t^{n-1}P_2(t,\alpha,\varepsilon)\,dt > -\infty$$

respectively.

This is actually a new result, but it can be obtained as a consequence of a result of the author in [134; Chapter 3]; it was stated under slightly less general hypotheses ($P_1 = -P_2$ bounded in t) by Onose in [243]. The proof is based on Schauder's fixed point theorem for the operator

$$(Tf)(t) = K + \int_t^{\infty}(t-s)^{n-1}H(s,f(s),f'(s),\ldots,f^{(n-1)}(s))\,ds/(n-1)!$$

(K a nonzero constant) in a suitable Banach space, and is omitted.

Another aspect of interest is the following: under what conditions on H does (I) have at least one oscillatory solution? This question was answered by Jasny [116] and Kurzweil [175] in the case $n = 2$. The reader is also referred to the papers of Heidel and Hinton [105], and Heidel and Kiguradge [106] for further second order considerations. The case n odd has been covered by the following result of Heidel.

THEOREM 2.11. (Heidel [103]). In the equation (I) assume that $n \geq 3$ is odd and that $H(t,u) \equiv p(t)u^{\gamma}$, where $p : R_T \to R_+$, $\gamma > 1$ is the quotient of two odd positive integers, and

$$\int_T^{\infty} s^{n-2+\gamma}p(s)\,ds = +\infty$$

Then (I) has at least one oscillatory solution which is not identically equal to zero on any infinite ray of $[T,\infty)$.

Concluding this section, we should mention that Lyapunov functions were introduced in oscillation theory by Yoshizawa in [348, 349]. Yoshizawa considered second order equations, and it would be very interesting to see some developments in the nth order case. Howard and Hayden [113] abstracting a criterion of Howard [110] of 1962, established several results for equations with values in a certain Banach operator space. Domšlak [64, 66] considered differential equations in Hilbert spaces and finite dimensional systems. Such systems were also considered by the author [134] and Olekhnik [237]. Finally, Nehari [234] studied a quasi-linear system and established an interesting oscillation property.

THE EQUATION (II)

The first result that ensured the oscillation of all solutions of a forced nonlinear equation (II) is contained in [136] and a simpler case than the one considered there could be stated as follows:

THEOREM 3.1. (Kartsatos [136]). Consider the equation (II) with $H(t,u) \equiv p(t)g(u)$ and Q_1 satisfying (Q_1), where $p : R_T \to R_+^o$, $g : R \to R$, g increasing, $ug(u) > 0$ for $u \neq 0$ and such that (2.14) is satisfied. Then if the function $P(t)$ $(P^{(n)}(t) \equiv Q_1(t))$ is oscillatory and such that $\lim_{t\to\infty} P(t) = 0$, (II) is oscillatory.

A corresponding result also holds for bounded solutions if all bounded solutions of (I) are assumed to be oscillatory. In 1972, the author answered a problem given by Wong in [342]. This problem concerns the oscillation of forced equations with periodic forcing term. A simpler version of it is the following:

THEOREM 3.2. (Kartsatos [138]). Let H be as in Theorem 3.1 without the integral conditions on g. Moreover, let Q_1 satisfy (Q_1) with $P(t)$ periodic and oscillatory on R_T. Then (II) is oscillatory if (I) is oscillatory.

Actually, a weaker result was proved in [138]. In the theorem above we took into consideration the comparison Theorem 2.6. Simultaneously, a paper of Teufel [319] appeared ensuring the oscillation of all solutions of (II) with $n = 2$ for certain forcing terms whose class also contains the periodic functions. We shall give a simplified version of Theorem 2 in [319].

THEOREM 3.3. (Teufel [319]). Let $n = 2$ in (II) and assume that $H(t,u) \equiv p(t)g(u)$ with $p : R_T \to R_+^0$, $g : R \to R$ is increasing, and $ug(u) > 0$ for $u \neq 0$. Moreover, let there exist two positive constants δ, Δ and a sequence of intervals of the form $[a + mb, a + mb + 3\Delta]$, $m = 1,2,\ldots$, on which the function $Q_1 : R_T \to R$ satisfies $Q_1(t) < -\delta$, and a sequence of intervals $[c + md, c + md + 3\Delta]$, $m = 1,2,\ldots$, on which $Q_1(t) > \delta$. Then the conditions

$$\int_T^\infty p(t)\,dt = +\infty \quad \text{and} \quad \left|\int_T^\infty Q_1(s)\,ds\right| = o\left(\int_T^t p(s)\,ds\right)$$

are sufficient for (II) to be oscillatory.

As Teufel notes in [319], Duffing's equation satisfies the above hypotheses.

Theorems 3.1 and 3.2 were extended by True [325; Chapter 3] to functional differential equations. Similar considerations can be found in the papers of True [326], Kusano and Onose [183] and Onose [247].

At the annual AMS meeting in 1973 (San Francisco), Professor Grimmer announced several results that can be obtained by looking at things in a quite different manner. These results concern homogeneous equations and are contained in [95]. They have been based on the fact that "a nonoscillatory solution of (I) satisfies a first order integral inequality while its (n-1)st derivative satisfies a first order differential inequality. By applying the comparison principle, results are obtained by analyzing the two associated first order scalar differential equations." These basic results of Grimmer were extended to forced equations by Foster [73, 74]. We are quoting here some of the results of Foster in [73].

THEOREM 3.4. (Foster [73]). Let (H) and (Q_1) hold with H increasing in x, and P taking positive and negative values on any interval $[b,\infty)$, $b \geq T$. Moreover, assume that for any $a > 0$ and $v_0 \neq 0$ all solutions $v(t,a,v_0)$ of the problem

$$v' = (t-a)^{n-1}H(t,v)/(n-1)!, \quad v(a) = v_0 \tag{3.1}$$

have finite escape times. Then if $\lim_{t\to\infty} P(t) = 0$, (II) is oscillatory.

THEOREM 3.5. (Foster [73]). Assume that the hypotheses of the above theorem are satisfied (except $P(t) \to 0$ as $t \to \infty$). Moreover, assume that there exist two sequences $\{s_m\}$, $\{t_m\}$ with s_m, $t_m \to \infty$ as $m \to \infty$, and such that $P(s_m) = M_0 > 0$, $P(t_m) = -L_0 < 0$, and $\inf_{s\geq a} P(s) = -L_0$, and $\sup_{s\geq a} r(s) = M_0$ for every $a > 0$. Then (II) is oscillatory.

THEOREM 3.6. (Foster [73]). Let the assumptions of Theorem 3.4 hold but with P(t) eventually of one sign. Then every solution of (II) either oscillates, or is eventually bounded between 0 and r(t) and tends to zero as $t \to \infty$.

These results of Foster, as well as those in [74] which cover the sublinear cases, are closely related to Theorem 3.1 and other results of the author in [140] (cf. Theorem 3.7 below); this is due to the existence of necessary and sufficient conditions for the oscillation of sublinear and superlinear cases when H(t,u) can be separated in the form p(t)g(u). The examples of Foster in [73, 74] are also covered by a result of the author in [140], which we now state.

THEOREM 3.7. (Kartsatos [140]). Let (H) and (Q_1) be satisfied with H increasing in u. Then under anyone of the following conditions (II) is oscillatory.

(i) P(t) is oscillatory and $P(t) \to 0$ as $t \to \infty$.

(ii) there exists $P_1(t)$ satisfying (Q_1) (with P replaced by P_1) and such that P and P_1 are oscillatory with $\liminf_{t\to\infty} P(t) = 0$ and $\limsup_{t\to\infty} P_1(t) = 0$.

This theorem extends to the nth order case Theorem 5 in Atkinson's paper [3], which in turn extends results of the author in [136] and [138]. The criterion (iii) in [3; Theorem 5] also holds in the nth order case; this follows from Theorem 3.3 of the author [140]. The reader should notice that no explicit integral conditions were assumed on H in Theorem 3.7. Thus, it provides a very general result in cases of "small" or "periodic-like" forcing terms. The part of the above theorem referring to (i) has been extended to functional equations by Onose [248]. Atkinson's paper [3] contains several other results that need to be extended to the nth order case (cf. also Kartsatos [140]).

Now denote by F_k the class of all n times continuously differentiable functions u(t), $t \in [T,\infty)$ such that $|u(t)| = O(t^k)$ as $t \to \infty$. Then we have the following.

THEOREM 3.8. (Staikos and Sficas [297]). Consider Equation (1.1) with $H : R_T \times R^n \to R$. Moreover, assume the existence of integers k, ℓ, and m such that $0 \leq m \leq n-1$, $0 \leq \ell \leq k \leq n-m-1$ and

$$\int_T^\infty t^m H(t,u(t),u'(t),\ldots,u^{(n-1)}(t))dt = \pm\infty \tag{3.2}$$

for every nonoscillatory $u \in F_k$ with $\liminf_{t\to\infty} (|u(t)|/t^\ell) \neq 0$. Then for all F_k - solutions x(t) of (1.1) we have $\liminf_{t\to\infty} |x(t)|/t^\ell = 0$.

This theorem implies that for k = 0 all bounded solutions x(t) of (1.1) are either oscillatory or such that $\liminf_{t\to\infty} |x(t)| = 0$. The case k = 0 and $0 \leq m \leq n-2$ was considered by the author in [136], which in turn extended Theorem 1 in [129] for second-order equations. For more consequences of the above theorem and for functional equations, the reader is referred to the same paper [297].

A result related to the above theorem is the following theorem of Graef and Spikes [87], which, besides being important in itself, implies an interesting corollary. Graef and Spikes considered functional equations. We take the "functional" $x(g(t)) \equiv x(t)$.

<u>THEOREM 3.9.</u> (Graef and Spikes [87]). Let (1.1) be such that

$$\int_T^\infty H(t,u(t),u'(t),\ldots,u^{(n-1)}(t))dt = +\infty \ (-\infty)$$

for any $u \in C^n[T,\infty)$ with $\liminf_{t\to\infty} u(t) > K$ ($\limsup_{t\to\infty} u(t) < -K$). Then every solution of (1.1) oscillates, or satisfies $\liminf_{t\to\infty} |x(t)| \leq K$.

For $K = 0$ this theorem answers a recent question raised by the author in [136]. As we mentioned above, the above theorem has the following very interesting corollary concerning forced equations.

<u>THEOREM 3.10.</u> (Graef and Spikes [87]). Assume that $H \equiv p(t)g(t,u,u',\ldots,u^{(n-1)}) - q(t,u,u',\ldots,u^{(n-1)})$ in (1.1) and moreover, assume that $p : R_T \to R_+^o$, g, $q : R_T \times R^n \to R$, $u_1 g(t,u_1,u_2,\ldots,u_n) > 0$ for $u_1 \neq 0$, g is bounded away from zero if u_1 is bounded away from zero, there exists $h : R_T \to R_+$ such that $|q(t,u_1,\ldots,u_n)| \leq h(t)$ for $(t,u_1,u_2,\ldots,u_n) \in R_T \times R^n$,

$$\int_T^\infty p(t)dt = +\infty, \text{ and } \lim_{t\to\infty} h(t)/p(t) = 0$$

Then every solution $x(t)$ of (1.1) either oscillates, or is such that $\liminf_{t\to\infty} |x(t)| = 0$.

Another corollary in [87] ensures that the conclusion above holds if $\lim_{t\to\infty} h(t)/p(t) = 0$ is replaced by

$$\int_T^\infty h(t)dt < +\infty$$

In the following result, which appears in [149], we assume that $Q_1(t)$ has an oscillating nth antiderivative which "stays away" from zero.

THEOREM 3.11. (Kartsatos and Manougian [149]). Assume that (H) is satisfied and there exists a continuous function $P(t)$ such that $P^{(n)}(t) \equiv Q_1(t)$, $t \in R_T$ and

$$\limsup_{t \to \infty} P(t) > 0, \quad \liminf_{t \to \infty} P(t) < 0$$

Moreover, assume that for some $k > 0$,

$$\int_T^\infty t^{n-1} |H(t, u(t) + P(t))| dt < +\infty$$

for all $u : R_T \to R$ with $|u(t)| \le k$, $t \in R_T$. Then (II) has at least one oscillatory solution.

An example of an equation for the above theorem is the following:

$$x^{(4)} + [1/(1+t^6)]x^{1/3} = \sin t$$

Here we can take $P(t) \equiv \sin t$. This equation has in fact infinitely many bounded solutions. This can be shown by use of Schauder's fixed point theorem in the equation

$$W(t) = \lambda + \int_t^\infty (t-s)^3 [1/(1+s^6)](W(s) + \cos s)^{1/3} ds/3!$$

This equation has a solution $W_\lambda(t)$ for any λ with $0 < \lambda < 1$, and the function $x_\lambda(t) \equiv W_\lambda(t) + \cos t$ satisfies $\lim_{t \to \infty} [x_\lambda(t) - \cos t] = \lambda$, and equation (II). Obviously, $x_\lambda(t)$, $0 < \lambda < 1$, is oscillatory.

The above result can be extended to all solutions of (II) (cf. remarks following the above theorem in [149]).

Now we are quoting a useful theorem from Graef and Spikes [87] which improves a result of Kartsatos and Manougian [149; Theorem 2.1].

<u>THEOREM 3.12.</u> (Graef and Spikes [87]). Consider the equation

$$x^{(n)} + H(t,x,x',\ldots,x^{(n-1)}) = Q_1(t) \tag{3.3}$$

where $H : R_T \times R^n \to R$, $Q_1 : R_T \to R$ and $x_1 H(t,u_1,\ldots,u_n) > 0$ for $u_1 \neq 0$. Moreover, assume for every $k > 0$ and $t \geq T$ we have

$$\limsup_{t \to \infty} \left[\int_{\bar{t}}^{t} (t-s)^{n-1} Q_1(s)\,ds - kt^{n-1}\right] > 0 \tag{3.4}$$

$$\liminf_{t \to \infty} \left[\int_{\bar{t}}^{t} (t-s)^{n-1} Q_1(s)\,ds + kt^{n-1}\right] < 0 \tag{3.5}$$

Then (3.3) is oscillatory.

Manougian and the author assumed in [149] that the second members of (3.4) and (3.5) were $+\infty$ and $-\infty$ respectively. It is important to notice here that no growth condition has been placed on the function H. Thus, the oscillation in the above theorem could be created by the forcing term $Q_1(t)$. In this connection, the following theorem was proved in [148].

<u>THEOREM 3.13.</u> (Kartsatos and Manougian [148]). Consider Equation (II) under hypotheses (H) and (Q_1). Moreover, let

$$\limsup_{t \to \infty} \int_T^t H(s, \lambda + P(s))\,ds = +\infty, \text{ and}$$

$$\liminf_{t \to \infty} \int_T^t H(s, -\lambda + P(s))\,ds = -\infty$$

for every $\lambda > 0$. Then (II) is oscillatory.

This theorem was actually shown in the separated case $H(t,u) \equiv p(t)g(u)$, but its proof carries over to the present case without modifications. An interesting corollary to the above result reads as follows.

<u>COROLLARY 3.14.</u> (Kartsatos and Manougian [148]). Let $H(t,u) \equiv p(t)u^{2\mu+1}$ with $p : R_T \to R_+$ and μ a nonnegative integer. Moreover, let (Q_1) be satisfied with P oscillatory,

$$\int_T^\infty p(t)\,dt = +\infty, \text{ and } \int_T^\infty p(t)|P(t)|^m dt < +\infty$$

for every $m = 1,2,\ldots,2\mu + 1$. Then (II) is oscillatory.

An example of Theorem 3.13 is the following equation

$$x^{(4)} + (1/t^5)x = [t^6 \sin t]^{(4)} \tag{3.6}$$

Here, $T = 1$, $P(t) = t^6 \sin t$ and $\int_1^t H(s, \pm\lambda + P(s))ds =$ $\pm\lambda\int_1^t (1/s^5)ds - t\cos t + \sin t + (\cos 1 - \sin 1)$. It is well known that the homogeneous equation does not oscillate (cf., for example, Onose [240]).

THE EQUATION (III)

In this section, several theorems are quoted from the author's papers [143, 144]. In [143], the author initiated the study of perturbed equations as far as oscillation is concerned. The main result in [143] concerns itself with the B-oscillation of (III) provided that the homogeneous equation (I) is B-oscillatory.

<u>THEOREM 4.1.</u> (Kartsatos [143]). Consider equation (III) with $|Q_2(t,u)| \le Q(t)|u|^r$, $r > 1$, where $Q : R_T \to R_+$ and such that

$$\int_T^\infty t^{n-1} Q(t)\,dt < +\infty \tag{4.1}$$

Moreover, let (H) be satisfied. Then (III) is B-oscillatory if (I) is B-oscillatory.

In the following corollary it is shown that Lipschitzian perturbations imply the "relative" oscillation of pairs of bounded solutions. This phenomenon should be further researched because it implies certain asymptotic behavior of one bounded solution as compared to another bounded, but known, solution.

<u>COROLLARY 4.2.</u> (Kartsatos [143]). Consider Equation (III) with $H(t,u) \equiv P(t)u$ with $P(t)$ positive, continuous, and let $B_p = \{x \in R;\ |x| \le p\}$. Let $Q(t,u)$ be continuous and such that

$|Q(t,u_1) - Q(t,u_2)| \leq Q_o(t,p)|u_1 - u_2|$ for every u_1, $u_2 \in B_p$. Here $Q_o : R_T \times (R_+ - \{0\}) \to R_+$ is continuous and such that

$$\int_T^\infty t^{n-1} Q_o(t,p)dt < +\infty$$

Then the difference $\Delta(t) = x(t) - y(t)$ oscillates, where $x(t)$, $y(t)$ are any two bounded solutions of (III).

The proof follows immediately from the above theorem because the function $\Delta(t)$ satisfies the equation $\Delta''(t) + P(t)\Delta(t) = Q(t,x(t)) - Q(t,y(t))$.

The following result provides a comparison of the non-oscillatory solutions of (III) to those of the equation $u^{(n)} = Q_o(t,u)$, where Q_o is as in (Q_2).

<u>THEOREM 4.3.</u> (Kartsatos [143]). Assume that H is as in (H) and (Q_2) is satisfied with $Q_o(t,u)$ increasing in u and such that $uQ_o(t,u) \geq 0$ for all $(t,u) \in R_T \times R$. Let $x(t)$ be a positive solution of (III). Then there exists a constant $M > 0$ and a point $\bar{t} \geq T$ such that $x(t) < y(t)$ for $t \geq \bar{t}$, where $y(t)$ is any solution of $v^{(n)} = Q_o(t,v)$ such that $v(\bar{t}) \geq M$ and $v^{(i)}(\bar{t}) = 0$, $i = 1,2,\ldots,n-1$.

We shall indicate below how the above theorem can be used to formulate criteria for oscillation of perturbed equations. Before we do this, we state a theorem from [144] which again ensures the B-oscillation of (III).

<u>THEOREM 4.4.</u> (Kartsatos [144]). Let H satisfy (H) and be increasing in u. Assume further that for every $\alpha > 0$ there exists a function $Q_\alpha : R_T \to R_+$ such that $|Q(t,u)| \leq Q_\alpha(t)$ for every $u \in R$ with $|u| \leq \alpha$; here $Q : R_T \times R \to R$. Finally, let

$$\int_T^\infty t^{n-1}[H(t,\pm k) \mp Q_\alpha(t)]dt = \pm\infty$$

for any $k > 0$ and $\alpha > 0$. Then if $x(t)$ is a bounded eventually positive (negative) solution of (III), there exists a sequence $\{t_n\} \to \infty$ such that $H(t_n,x(t_n) \leq Q(t_n,x(t_n))(H(t_n,x(t_n)) \geq Q(t_n,x(t_n)))$. If, in addition, we assume that

$$\limsup_{t \to \infty} \{|Q(t,u)/H(t,u)| : |u| \leq k\} = 0$$

then (III) is B-oscillatory.

Now, for any $\bar{t} \geq T$ and any $M > 0$, let $C(\bar{t},M) = \{u \in C[\bar{t},\infty) : 0 < u(t) \leq y(t,M),\ t \in [\bar{t},\infty)\}$, where $y = y(t,M)$ is as in Theorem 4.3 but fixed (for each $M > 0$) and such that $y(\bar{t},M) = M$. Then we have the following theorem which actually contains as a special case both Theorems 3 and 4 of the author in [144]; the proofs there carry over to this case without modifications.

THEOREM 4.5. (Kartsatos [144]). Let the assumptions of Theorem 4.3 be satisfied and let $h : R \to R$ be increasing and such that $uh(u) > 0$ for $u \neq 0$. Furthermore, assume that for each $\bar{t} \geq T$, $M > 0$, and $u \in C(\bar{t},M)$, the equation

$$x^{(n)} + [H(t,u(t)) - Q(t,u(t))]/h(u(t))]h(x) = 0$$

is oscillatory. Then the first conclusion of Theorem 4.4 holds for all solutions of (III). If, moreover,

$$\lim_{t\to\infty}[Q(t,u(t))/H(t,u(t))] = 0$$

for every $\bar{t} \geq T$, $M > 0$ and $u \in C(\bar{t}, L)$, then every solution of (I) is oscillatory.

COROLLARY 4.6. (Kartsatos [144]). Consider the equation

$$x'' + p(t)|x|^{\alpha}\operatorname{sgn} x = q(t)|x|^{\beta}\operatorname{sgn} x$$

where p and q are positive for $t \geq 0$, and $0 < \alpha < \beta < 1$; furthermore, suppose that

$$\int_0^\infty t^{\alpha}p(t)dt = +\infty, \quad \int_{\bar{t}}^\infty [p(t) - q(t)(u(t))^{\beta-\alpha}]dt = +\infty$$

and

$$\lim_{t\to\infty}[q(t)(u(t))^{\beta-\alpha}/p(t)] = 0$$

for every $\bar{t} \geq 0$ and $M > 0$, where $u(t)$ is any function with

$$0 < u(t) \leq [M^{1-\beta} + (1-\beta)\int_{\bar{t}}^{t}(t-s)q(s)ds]^{1/(1-\beta)}$$

for $t \geq \bar{t}$. Then (4.2) is oscillatory.

The last assertion in Theorems 4.4, 4.5, and Corollary 4.6 can be shown independently by use of the comparison Theorem 2.6. As an example of the above corollary, we can take $\alpha = 1/5$, $\beta = 1/3$, $p(t) = (t+1)^{-1}$, $q(t) = (t+1)^{-3}$. Then $u(t) \leq \lambda t^{1/2}$ (for some $\lambda > 0$) and the second of the above integral conditions becomes

$$\int_{\bar{t}}^{\infty} t^{1/5}[(t+1)^{-1} - \lambda(t+1)^{-3}t^{1/2}]dt = +\infty$$

which is true for all $\lambda > 0$ and $t \geq 0$.

SOME NEW RESULTS

We consider first a differential equation of the form

$$x^{(n)} + p(t)x^{(n-1)} + H(t,x) = Q(t) \tag{5.1}$$

where H satisfies (H), $p : R_T \to R$ and $Q : R_T \to R$. Moreover by $r(t)$ we denote a fixed solution of the equation

$$r^{(n)}(t) + p(t)r^{(n-1)}(t) = Q(t) \tag{5.2}$$

for $t \geq T$. Then if we let $w(t) \equiv x(t) - r(t)$, where $x(t)$ is a solution of (5.1), we obtain

$$w^{(n)}(t) + p(t)w^{(n-1)} + H(t,w + r(t)) = 0 \tag{5.3}$$

Now we present a lemma which was given by Onose and the author in [150] and, for the sake of completeness, we include a rather simpler proof of it.

<u>LEMMA 5.1.</u> (Kartsatos and Onose [150]). Consider (5.1) with $Q \equiv 0$. Moreover, let $p(t) \leq m(t)$ where $m : R_T \to R_+$ satisfies

$$\lim_{t\to\infty}\int_{\bar{t}}^{t}\exp[-\int_{\bar{t}}^{u}m(s)ds]du = +\infty$$

for every $\bar{t} \geq T$. Then if $x(t)$ is a nonoscillatory solution of (5.1), we have $x^{(n-1)}(t)x(t) > 0$ for all large t.

<u>Proof.</u> Let $x(t)$ be a nonoscillatory solution of (5.1) and assume that $x(t) > 0$, $t \geq t_1 \geq T$. Now let $x^{(n-1)}(t_o) = 0$ for some $t_o \geq t_1$. Then

$$x^{(n)}(t_o) = -H(t_o, x(t_o)) < 0 \tag{5.4}$$

which implies that $x^{(n-1)}(t)$ cannot have another zero after it vanishes once. Thus, $x^{(n-1)}(t)$ has fixed sign for all large t. Let $x^{(n-1)}(t) < 0$ for all large t (say $t \geq t_2 \geq t_1$). Then if we put $\phi(t) \equiv x^{(n)}(t) + p(t)x^{(n-1)}(t)$, $t \geq t_2$, we get

$$x^{(n)}(t) + p(t)x^{(n-1)}(t) = \Phi(t) = -H(t, x(t)) < 0, \tag{5.5}$$

for $t \geq t_2$. Thus, solving for $x^{(n-1)}(t)$, we find

$$\begin{aligned} x^{(n-1)}(t) &= \exp\left[-\int_{t_2}^{t} p(s)\,ds\right]\left[x^{(n-1)}(t_2) + \int_{t_2}^{t} \phi(u)\exp\left[\int_{t_2}^{u} p(s)\,ds\right]du\right] \\ &\leq x^{(n-1)}(t_2)\exp\left[-\int_{t_2}^{t} p(s)\,ds\right] \\ &\leq x^{(n-1)}(t_2)\exp\left[-\int_{t_2}^{t} m(s)\,ds\right] \end{aligned} \tag{5.6}$$

An integration of (5.6) from t_2 to $t \geq t_2$ yields $x^{(n-2)}(t) \to -\infty$ as $t \to +\infty$, which implies $\lim_{t\to\infty} x(t) = -\infty$, a contradiction. Thus $x^{(n-1)}(t)x(t) > 0$ for $x(t)$ eventually positive, and a similar proof can be given for an eventually negative $x(t)$.

The above lemma has been also shown by Naito [228] by use of Langenhop's inequality, and part of its proof for n=2 goes back to Bobisud [23].

THEOREM 5.2. Let $p(t)$ in (5.1) be nonnegative for $t \geq T$ and assume that $\limsup_{t \to \infty} r(t) = +\infty$, $\liminf_{t \to \infty} r(t) = -\infty$. Then (5.1) is B-oscillatory.

Proof. Let $x(t)$, $t \geq t_1 \geq T$ be a bounded positive solution of (5.1). Then $w(t) \equiv x(t) - r(t)$, $t \geq t_1$, satisfies (5.3). Since $w(t) + r(t) > 0$ for $t \geq t_1$, it follows from the proof of Lemma 5.1 that $w^{(n-1)}(t) > 0$, or $w^{(n-1)}(t) < 0$ for all large t. Thus, $w(t)$ is monotonic (and hence of one sign) for all large t. Consequently, $x(t) - r(t) > 0$, or $x(t) - r(t) < 0$ for all large t. This implies $\limsup_{t \to \infty} x(t) = +\infty$ or $\liminf_{t \to \infty} x(t) = -\infty$ respectively. Since $x(t)$ is bounded, we have the desired contradiction. A similar proof holds for $x(t)$ eventually negative.

It is interesting to remark here that no growth condition was placed on the function H in the above theorem.

THEOREM 5.3. Let p and H be as in Theorem 5.2 and Lemma 5.1, and assume that $r(t)$ is oscillatory and satisfies $\lim_{t \to \infty} r(t) = 0$. Then if (I) is oscillatory (B-oscillatory), (5.1) is oscillatory (B-oscillatory).

Proof. Let $x(t)$ be an eventually positive solution of (5.1). Then as in the proof of Lemma 5.1, $w^{(n-1)}(t)w(t) > 0$ for all large t, say for $t \geq t_1 \geq T$. Let $w(t) > 0$ for $t \geq t_1$. Since n is even, it follows from usual arguments that $w'(t) > 0$ for (say) $t \geq t_2 \geq t_1$. Now let ε be given with $w(t_2) > \varepsilon > 0$, and let $|r(t)| < \varepsilon$ for $t \geq t_2$. Thus $w(t) + r(t) \geq w(t) - \varepsilon > 0$ for $t \geq t_2$. Consequently, from (5.3) we obtain

$$w^{(n)}(t) + H(t,w(t) - \varepsilon)$$
$$\leq w^{(n)}(t) + H(t,w(t) + r(t)) \leq 0 \tag{5.7}$$

for every $t \geq t_2$. If we let $w(t) - \varepsilon \equiv v(t)$, $t \geq t_2$, then it follows from (5.7) that the inequality

$$y^{(n)}(t) + H(t,y(t)) \leq 0 \qquad (5.8)$$

has a positive solution $v(t)$, $t \geq t_2$. Now we can use Lemma 2.1 in [140], where it is concluded that (I) must also have a positive solution on $[t_2,\infty)$. This leads to a contradiction if we assume that (I) is oscillatory. The same conclusion can be drawn if we assume that (I) is B-oscillatory by calling our attention to Corollary 2.1 in [140]. Thus, $x(t) < r(t)$ for all large t which is a contradiction to the positiveness of $x(t)$. A similar argument covers the case of an eventually negative $x(t)$, and this completes the proof.

Actually, as in Theorem 3.7, we could have assumed instead of $\lim_{t\to\infty} r(t) = 0$, the existence of another function $r_1(t)$ such that $r_1(t)$ is also oscillatory, $r_1^{(n)}(t) + p(t)r_1^{(n-1)}(t) = Q(t)$, r, r_1 bounded, $\limsup_{t\to\infty} r_1(t) = 0$, and $\liminf_{t\to\infty} r(t) = 0$.

The above theorem improves Theorems 1 and 2 of Naito's paper [228], and it also includes the linear case. It should be remarked now that if we multiply (5.1) by $s(t) \equiv \exp[\lambda(t)]$, where $\lambda(t)$ is any antiderivative of $p(t)$, we obtain

$$[s(t)x^{(n-1)}]' + H_1(t,x) = Q_3(t) \qquad (5.9)$$

where $H_1(t,u) \equiv s(t)H(t,u)$ and $Q_3(t) \equiv s(t)Q(t)$. In this case, if we let $r(t)$ satisfy

$$[s(t)r^{(n-1)}(t)]' = Q_3(t) \qquad (5.10)$$

we obtain, instead of (5.3),

$$[s(t)w^{(n-1)}]' + H_1(t,w + r(t)) = 0 \qquad (5.11)$$

Let us now give an easy but important lemma concerning the monotonicity of nonoscillatory solutions of (5.11).

<u>LEMMA 5.4.</u> Consider (5.11), where $r(t)$ satisfies (5.10) and is bounded, and $w(t) \equiv x(t) - r(t)$ corresponds to a positive (negative) solution $x(t)$ of (5.9) with $H(t,x)$ satisfying (H). Moreover, let

$$\int_T^\infty [s(u)]^{-1} du = +\infty$$

Then $w^{(n-1)}(t)w(t) > 0$ for all large t.

Proof. Let $w(t) + r(t) > 0$, $t \geq t_1 \geq T$. Then since $[s(t)w^{(n-1)}(t)]' < 0$, $t \geq t_1$, $s(t)w^{(n-1)}(t)$ is strictly decreasing. Let $t_2 \geq t_1$ be such that $s(t)w^{(n-1)}(t) < 0$ for $t \geq t_2$. Then

$$s(t)w^{(n-2)}(t) < s(t_2)w^{(n-2)}(t_2) = \mu < 0 \tag{5.12}$$

for every $t \geq t_2$. Dividing by $s(t)$ and integrating we get $\lim_{t\to\infty} w^{(n-2)}(t) = -\infty$. This implies that $\lim_{t\to\infty} w(t) = -\infty$ which contradicts the fact that $w(t) + r(t) > 0$ and $r(t)$ is bounded. Consequently, $w^{(n-1)}(t) > 0$ for all large t, and similar considerations cover the case of a negative $x(t)$. This completes the proof.

Now it is easy to state criteria for the oscillation of (5.11) by taking into consideration the above lemma. We omit the corresponding statements which resemble those of the above theorems.

The reader is also referred to the paper of Sficas [269] for some developments concerning functional equations.

Let us now make some remarks about the case

$$x^{(n)} + p(t)x^{(n-2)} + H(t,x) = Q(t) \tag{5.12}$$

Let $x(t)$ be a solution of (5.12) and $w(t) \equiv x(t) - r(t)$, where $r(t)$ is a solution of $r^{(n)} + p(t)r^{(n-2)} = Q(t)$. Then $w(t)$ satisfies

$$w^{(n)} + p(t)w^{(n-2)} + H(t,w + r(t)) = 0 \tag{5.13}$$

Now we can state the following lemma concerning the sign of $w^{(n-2)}(t)$ in (5.13).

LEMMA 5.5. In (5.13) assume that $H : R_T \times R \to R$, $p : R_T \to R_-$, and $Q : R_T \to R$ with $uH(t,u) < 0$ for every $(t,u) \in R_T \times (R - \{0\})$. If $w(t) + r(t) > 0$ ($w(t) + r(t) < 0$) for $t \geq t_1 \geq T$, then $w^{(n-2)}(t)$ is of fixed sign on the interval $[t_1,\infty)$. If, moreover,

$$\int_T^{\infty} tp(t)\,dt > -\infty \tag{5.14}$$

then $w^{(n-2)}(t) > 0$ ($w^{(n-2)}(t) < 0$) eventually.

Proof. Let $w(t) + r(t) > 0$ for $t \geq t_1 \geq T$. Moreover, assume that $y(t) \equiv w^{(n-2)}(t)$ takes positive and negative values for all large t. Then y(t) satisfies the equation

$$y'' + p(t)y + q(t) = 0$$

where $q(t) \equiv H(t,w(t) + r(t)) < 0$, $t \geq t_1$. Consequently, in each interval of positiveness of y(t), we have $y''(t) > 0$. This implies that y(t) is convex whenever it is positive, and this contradicts the fact that it is oscillatory. Thus, y(t) is of fixed sign for all large t. The rest of the proof follows exactly as in Theorem 3.12 of Liossatos [205] and we omit it.

The dissertation of Liossatos [205] is devoted to the equation (5.12) with $Q(t) \equiv 0$, and concerns itself with the extension of the main results of Heidel's paper [101] to the general nth order case.

We note now that the integral condition on p(t) in the above lemma can be replaced by the condition $p(t) \geq -2/t^2$, $t \geq T$, (cf. Liossatos [205; Theorem 3.1.3]).

In the following theorem conditions are given for the B-oscillation of equation (5.12).

THEOREM 5.6. Let p, H, and Q be as in Lemma 5.5, and let p(t) satisfy (5.14). Moreover, let $r_1(t)$ and $r_2(t)$ be oscillatory with $r_i^{(n)} + p(t)r_i^{(n-2)} = Q(t)$, $i = 1,2$, $\liminf_{t\to\infty} r_1(t) = 0$, $\limsup_{t\to\infty} r_1(t) = +\infty$, $\liminf_{t\to\infty} r_2(t) = -\infty$, and $\limsup_{t\to\infty} r_2(t) = 0$. Then (5.12) is B-oscillatory.

Proof. Let $x(t)$ be a bounded nonoscillatory solution of (5.12) and assume that $x(t) > 0$ for $t \geq t_1 \geq T$. Let $w(t) \equiv x(t) - r_1(t)$. Then $w(t)$ satisfies (5.12) (with r replaced by r_1) for every $t \geq t_1$. Since $w(t) + r_1(t) > 0$, it follows from Lemma 5.5 that $w^{(n-2)}(t) > 0$ for every $t \geq t_2$ for some $t_2 \geq t_1$. Then $w(t)$ is either positive or negative for all large t. This implies $x(t) > r_1(t)$ or $x(t) < r_1(t)$ for all large t, which both yield a contradiction. We omit the argument for a negative $x(t)$.

The following theorem improves a result of Liossatos [205; Theorem 3.1.4].

THEOREM 5.7. Let p, H, and Q be as in Lemma 5.5, let $p(t)$ satisfy (5.14), H be decreasing in u, and

$$\int_T^{\infty} t^{n-1} H(t, \pm k)\,dt = \mp\infty \tag{5.15}$$

for every $k > 0$. Then if $r(t)$ is oscillatory and $\lim_{t\to\infty} r(t) = 0$, every nonoscillatory solution $x(t)$ of (5.12) satisfies $\lim_{t\to\infty} |x(t)| = 0$ or $+\infty$.

Proof. Let $x(t)$ be a solution of (5.12) with $x(t) > 0$ for $t \geq t_1 \geq T$. Let $w(t) \equiv x(t) - r(t)$. Since $w(t) + r(t) > 0$, it follows from Lemma 5.5 that $w^{(n-2)}(t) > 0$ for every $t \geq t_2$ for some $t_2 \geq t_1$. Thus $w(t)$ is either positive or negative and monotonic for all large t. Let $w(t) > 0$ for all $t \geq t_3 \geq t_2$. Then if $w(t)$ is unbounded, it satisfies $\lim_{t\to\infty} w(t) = +\infty$, which implies $\lim_{t\to\infty} x(t) = +\infty$. If $w(t)$ is bounded, then, since n is even, $w'(t) < 0$ for $t \geq t_3$. Actually, we have $(-1)^j w^{(j)}(t) > 0$ for $j = 1, 2, \ldots, n$. This follows because no two consecutive derivatives of $w(t)$ can eventually be of the same sign due to the boundedness of $w(t)$. Now let $\lim_{t\to\infty} w(t) = A > 0$. Then, given ε with $0 < \varepsilon < A/2$ there exists $t_4 \geq t_3$ such that

$w(t) \geq A - \varepsilon$ and $r(t) \geq -\varepsilon$ for all $t \geq t_4$. Thus, $w(t) + r(t) \geq A - 2\varepsilon > 0$ for all $t \geq t_4$. Now consider the function $t^{n-1}x^{(n-1)}(t)$. By differentiation of this function and then integration from t_4 to $t \geq t_4$ we obtain

$$t^{n-1}w^{(n-1)}(t) - (n-1)\int_{t_4}^{t} s^{n-2}w^{(n-1)}(s)\,ds$$

$$\geq t_4^{\,n-1}w^{(n-1)}(t_4) - \int_{t_4}^{t} s^{n-1}H(s,A-2\varepsilon)\,ds$$

which implies, because of the negativeness of $w^{(n-1)}(t)$ and the integral condition on H,

$$\lim_{t\to\infty}\int_{t_4}^{t} s^{n-2}w^{(n-1)}(s)\,ds = -\infty \tag{5.16}$$

Now the proof follows as in Theorem 1 of [133] to obtain $\lim_{t\to\infty} w(t) = +\infty$, a contradiction to the boundedness of $w(t)$. Consequently, $A = 0$, in which case $\lim_{t\to\infty} x(t) - r(t) = 0 = \lim_{t\to\infty} x(t)$. It remains to consider $w(t) < 0$ for all large t, but this is not allowed due to the oscillation of $r(t)$. The analogous proof for $x(t)$ eventually negative is omitted.

Liossatos assumed in [205] that $H(t,u) = q(t)g(u)$, with $q(t) \leq 0$, $ug(u) > 0$ for $u \neq 0$, $Q(t) \equiv 0$, and

$$\int_{T}^{\infty} t^2 q(t)\,dt = -\infty \tag{5.17}$$

instead of (5.15). We should note here that if H is separated as above, then $g(u)$ does not have to be increasing for all u, as it would follow from the assumption on H of the above theorem.

It is now obvious that if we want to apply the methods exhibited above we must have qualitative information about the solutions of (5.2) and (5.3) which is a serious problem in itself. If we try functions $r(t)$ with $r^{(n)}(t) = Q(t)$, then the equations (5.3) and (5.13) must be replaced by

$$w^{(n)} + p(t)[w + r(t)]^{(n-1)} + H(t,w + r(t)) = 0 \tag{5.18}$$

$$w^{(n)} + p(t)[w + r(t)]^{(n-2)} + H(t,w + r(t)) = 0 \tag{5.19}$$

for which there is nothing known (cf. Problems IV and V below).

OPEN PROBLEMS

PROBLEM I. Extend Atkinson's Theorem 2.2 to nth order equations.

PROBLEM II. Extend and improve (by considering linear, superlinear, and sublinear cases) Teufel's Theorem 3.3 to nth order equations.

PROBLEM III. Extend Heidel's Theorem 2.11 to even order equations.

PROBLEM IV. Study the oscillation of (5.1) using (5.18) where $w = x - r$, $r^{(n)} = Q$.

PROBLEM V. Study the oscillation of (5.12) using (5.19) where $w = x - r$, $r^{(n)} = Q$.

PROBLEM VI. Provide conditions for the oscillation of (5.12) in case n even, $p(t) \leq 0$, and $uH(t,u) < 0$ for $u \neq 0$. (None known).

PROBLEM VII. Study the behavior of (5.12) where p and H are not necessarily as above.

PROBLEM VIII. Extend Corollary 4.2 to equations with nonlinear homogeneous parts.

PROBLEM IX. Establish conditions under which (III) is oscillatory without necessarily requiring

$$\lim_{t\to\infty}[Q(t,u(t))/H(t,u(t))] = 0$$

as in Theorem 4.5.

PROBLEM X. Same as above but for bounded solutions without necessarily assuming that

$$\limsup_{t\to\infty} \sup_{|u|\leq k} [Q(t,u)/H(t,u)] = 0$$

as in Theorem 4.4.

PROBLEM XI. Extend the comparison Theorem 2.6 in any possible direction.

PROBLEM XII. Study nth order forced "dynamical systems" generalizing results of Graef [80] and Burton, Townsend [37].

PROBLEM XIII. Prove or disprove: (II) is oscillatory if $H(t,u) \equiv p(t)g(u)$, $p : R_T \to R$, $g'(u) \geq 0$, $ug(u) > 0$ for $u = 0$, and, for every $\varepsilon > 0$,

$$\int_T^\infty t^{n-1} p(t)\,dt = +\infty, \quad \int_\varepsilon^\infty \frac{ds}{g(s)} < +\infty, \quad \int_{-\varepsilon}^{-\infty} \frac{ds}{g(s)} < +\infty$$

provided that Q_1 satisfies (Q_1) with $P(t) \to 0$ and oscillatory. (cf. Legatos and Kartsatos [200], Bobisud [22], Travis [323], Coles [55], Onose [246]). Consider also the general problem for the two other cases as in Problem II, as well as the case $Q_1 \equiv 0$.

PROBLEM XIV. Develop oscillation criteria in the spirit of Howard [112] for the equation (II). (Howard's results are rather difficult to apply, but very interesting and need some "smoothing". His methods go back to 1962 [110] (cf. also [111]. It seems that Condition 3 in Theorem 1 there is impossible)).

PROBLEM XV. Obtain a classification of solutions of (II) as in Ladas, Lakshmikantham, and Papadakis [196], by use of $w = x - P$ with $P^{(n)} = Q_1$ (cf. also Bogar [28], Staikos and Sficas [296], Liu [207]).

PROBLEM XVI. Find upper and lower bounds for the non-oscillatory solutions of (II) with $uH(t,u) > 0$. These bounds define classes of functions by which we can provide integral criteria for oscillation as in Theorem 4.5.

REFERENCES

1. G. V. Ananeva and B. I. Balaganskii, Oscillation of the solutions of certain differential equations of high order, Uspehi Mat. Nauk 14 (1959), 135-140.

2. F. V. Atkinson, On second order nonlinear oscillation, Pacific J. Math. 5 (1955), 643-647.

3. F. V. Atkinson, On second order differential inequalities, Proc. Royal Soc. Edinburgh Sect. A 72 (1972/73), 109-127.

4. J. W. Baker, Oscillation theorems of a second order damped nonlinear differential equation, SIAM J. Appl. Math. 25 (1973), 37-40.

5. R. Bellman, Stability theory of differential equations, McGraw-Hill, New York, 1953.

6. Š. Belohorec, Oscillatory solutions of certain nonlinear differential equations of second order, Mat.-Fyz. Časopis Sloven, Akad. Vied. 11 (1961), 250-255.

7. Š. Belohorec, Oscillatory solutions of certain nonlinear differential equations of second order, Mat.-Fyz. Časopis Sloven. Akad. Vied. 12 (1962), 253-262.

8. Š. Belohorec, On some properties of the equation $y''(x) + f(x)y^{\alpha}(x) = 0$, $0<\alpha<1$, Mat.-Fyz. Časopis Sloven. Akad. Vied. 17 (1967), 10-19.

9. Š. Belohorec, Monotone and oscillatory solutions of a class of nonlinear differential equations, Mat.-Fyz. Časopis Sloven. Akad. Vied. 19 (1969), 169-187.

10. Š. Belohorec, A criterion for oscillation and non-oscillation, Acta Fac. Rerum. Natur. Univ. Comenian 20 (1969-70), 75-79.

11. Š. Belohorec, Two remarks on the properties of solutions of a nonlinear differential equation, Acta Fac. Rerum. Natur. Univ. Comenian 22 (1969), 19-26.

12. D. C. Benson, Comparison theory for $y'^2 - 2p(x)y = -p(x)q(x)$, SIAM J. Appl. Math. 21 (1971), 279-286.

13. D. C. Benson, Oscillation of solutions of a generalized Liénard equation, Proc. Amer. Math. Soc. 33 (1972), 101-106.

14. D. C. Benson, Comparison and oscillation theory for Liénard's equation with positive damping, SIAM J. Appl. Math. 24 (1973), 251-271.

15. S. R. Bernfeld and A. Lasota, Quickly oscillating solutions of autonomous ordinary differential equations, Proc. Amer. Math. Soc. 30 (1971), 519-526.

16. S. R. Bernfeld and J. A. Yorke, The behaviour of oscillatory solutions of $x''(t)+p(t)g(x(t)) = 0$, SIAM J. Math. Anal. 3 (1972), 654-667.

17. N. P. Bhatia, Some oscillation theorems for second order differential equations, J. Math. Anal. Appl. 15 (1966), 442-446.

18. I. Bihari, Ausdehnung der Sturmschen Oszillations und Vergleichssätze auf die Lösungen gewisser nichtlinearen differentialgleichungen sweiter Ordnung, Magyar Tud. Acad. Mat. Kutató Int. Kozl. 2 (1957), 155-173.

19. I. Bihari, Extension of certain theorems of the Sturmian type to nonlinear second order equations, Magyar Tud. Acad. Mat. Kutató Int. Kozl. 3 (1958), 13-20.

20. I. Bihari, Oscillation and monotonity theorems concerning nonlinear differential equations of the second order, Acta Math. Sci. Hungarica 9 (1958), 83-104.

21. I. Bihari, An oscillation theorem concerning the half-line differential equation of the second order, Magyar Tud. Akad. Kutató Int. Kozl. 8 (1963), 275-280.

22. L. E. Bobisud, Oscillation of nonlinear second order equations, Proc. Amer. Math. Soc. 23 (1969), 501-505.

23. L. E. Bobisud, Oscillation of nonlinear differential equations with small nonlinear damping, SIAM J. Appl. Math. 18 (1970), 74-76.

24. L. E. Bobisud, Comparison and oscillation theorems for nonlinear second-order differential equations, J. Math. Anal. Appl. 32 (1970), 5-14.

25. L. E. Bobisud, Oscillation of solutions of damped nonlinear equations, SIAM J. Appl. Math. 19 (1970), 601-606.

26. L. E. Bobisud, Oscillation of second-order differential equations with retarded argument, J. Math. Anal. Appl. 43 (1973), 201-205,

27. G. A. Bogar, Oscillation properties of two term linear differential equations, Trans. Amer. Math. Soc. 161 (1971), 25-33.

28. G. A. Bogar, Oscillations of nth order differential equations with retarded argument, SIAM J. Math. Anal. 5 (1974), 473-481.

29. J. S. Bradley, Oscillation theorems for a second order delay equation, J. Differential Equations 8 (1970), 394-403.

30. F. Burkowski, Oscillation theorems for a second order nonlinear functional differential equation, J. Math. Anal. Appl. 33 (1971), 258-262.

31. F. Burkowski, Nonlinear oscillation of a second order sublinear functional differential equation, SIAM J. Appl. Math. 21 (1971), 486-490.

32. T. A. Burton and R. C. Grimmer, Stability properties of $(r(t)u')' + a(t)f(u)g(u') = 0$, Monatsh. Math. 74 (1970), 211-222.

33. T. A. Burton and R. Grimmer, On continuability of solutions of second order differential equations, Proc. Amer. Math. Soc. 29 (1971), 277-283.

34. T. A. Burton and R. Grimmer, On the asymptotic behavior of solutions of $x''+a(t)f(x) = e(t)$, Pacific J. Math. 41 (1972), 43-55.

35. T. Burton and R. Grimmer, Oscillatory solutions of $x'' + a(t)f(x(q(t))) = 0$, in Delay and Functional Diff. Equations and their Applications, Academic, New York, 1972, 335-343.

36. T. Burton and R. Grimmer, Oscillation, continuation, and uniqueness of solutions of retarded differential equations, Trans. Amer. Math. Soc. 179 (1973), 193-209.

37. T. A. Burton and C. G. Townsend, On the Generalized Liénard equation with forcing function, J. Differential Equations 4 (1968), 620-633.

38. G. J. Butler, The oscillatory behaviour of a second order nonlinear differential equation with damping, J. Math. Anal. Appl., to appear.

39. G. J. Butler, Oscillation theorems for a nonlinear analogue of Hill's equation, Quart. J. Math. Oxford, to appear.

40. G. J. Butler, The existence of continuable solutions of a second order differential equation, to appear.

41. Y. V. Bykov, L. Y. Bykova and E. I. Šercov, Sufficient conditions for oscillation of solutions of nonlinear differential equations with deviating argument, Differencial'nye Uravnenija 9 (1973), 1555-1560.

42. Y. V. Bykov and G. D. Merzlyakova, On the oscillatoriness of solutions of nonlinear differential equations with a deviating argument, Differencial'nye Uravnenija 10 (1974), 210-220.

43. Y. M. Chen, An oscillation criterion for the second order nonlinear differential equation $x'' + F(x'^2,x^2,t) = 0$, Quart. J. Math. Oxford Ser. (2) 24 (1973), 165-168.

44. Kuo-Liang Chiou, A second order nonlinear oscillation theorem, SIAM J. Appl. Math. 21 (1971), 221-224.

45. Kuo-Liang Chiou, A nonoscillation theorem for the superlinear case of second order differential equations $y'' + yF(y^2,x) = 0$, SIAM J. Appl. Math. 23 (1972), 456-459.

46. Kuo-Liang Chiou, The existence of oscillatory solutions for the equation $y''+q(t)y^r = 0$, $0<r<1$, Proc. Amer. Math. Soc. 35 (1972), 120-122.

47. Kuo-Liang Chiou, Oscillation and nonoscillation theorems for second order functional differential equations, J. Math. Anal. Appl. 45 (1974), 382-403.

48. C. V. Coffman and D. F. Ullrich, On the continuation of solutions of a certain nonlinear differential equation, Monatsh. Math. 71 (1967), 385-392.

49. C. V. Coffman and J. S. W. Wong, Second order nonlinear oscillations, Bull. Amer. Math. Soc. 75 (1969), 1379-1382.

50. C. V. Coffman and J. S. W. Wong, On a second order nonlinear oscillation problem, Trans. Amer. Math. Soc. 147 (1970), 357-366.

51. C. V. Coffman and J. S. W. Wong, Oscillation and nonoscillation of solutions of generalized Emden-Fowler equations, Trans. Amer. Math. Soc. 167 (1972), 399-434.

52. C. V. Coffman and J. S. W. Wong, Oscillation and nonoscillation theorems for second order ordinary differential equations, Funkcial. Ekvac. 15 (1972), 119-130.

53. W. J. Coles, A simple proof of a well-known oscillation theorem, Proc. Amer. Math. Soc. 19 (1968), 507.

54. W. J. Coles, An oscillation criterion for second-order linear differential equations, Proc. Amer. Math. Soc. 19 (1968), 755-759.

55. W. J. Coles, Oscillation criterion for nonlinear second order equations, Ann. Mat. Pura Appl. (4) 82 (1969), 123-133.

56. W. J. Coles, A nonlinear oscillation theorem, International Conference on Differential Equations, Academic, New York, 1975, 193-202.

57. W. J. Coles and D. Willett, Summability criteria for oscillation of second order linear differential equations, Ann. Mat. Pura Appl. (4) 79 (1968), 391-398.

58. R. S. Dahiya, Nonoscillation of arbitrary order retarded differential equations of non-homogeneous type, Bull. Austral. Math. Soc. 10 (1974), 453-458.

59. R. S. Dahiya, Oscillation generating delay terms in even order retarded equations, J. Math. Anal. Appl. 49 (1975), 158-164.

60. R. S. Dahiya and B. Singh, A Lyapunov inequality and nonoscillation theorem for a second order nonlinear differential-difference equation, J. Math. Phys. Sci. 7 (1973), 163-170.

61. R. S. Dahiya and B. Singh, On oscillatory behaviour of even order delay equations, J. Math. Anal. Appl. 42 (1973), 183-190.

62. R. S. Dahiya and B. Singh, Certain results on nonoscillation and asymptotic nature of delay equations, Hiroshima Math. J. 5 (1975), 7-15.

63. K. M. Das, Properties of solutions of certain nonlinear differential equations, J. Math. Anal. Appl. 8 (1964), 445-451.

64. Y. I. Domšlak, On the oscillation of solutions of vector differential equations, Soviet Math. Dokl. 11 (1970), 21-23.

65. Y. I. Domšlak, On oscillation of solutions of differential equations of second order, Differencial'nye Uravnenija 7 (1971), 205-214.

66. Y. I. Domšlak, On oscillatory and nonoscillatory solutions of vector differential equations, Differencial'nye Uravnenija 7 (1971), 961-969.

67. L. Erbe, Nonoscillatory solutions of second order non-linear differential equations, Pacific J. Math. 28 (1969), 77-85.

68. L. Erbe, Oscillation theorems for second order nonlinear differential equations, Proc. Amer. Math. Soc. 24 (1970), 811-814.

69. L. Erbe, Oscillation criteria for second order nonlinear differential equations, Ann. Mat. Pura Appl. (4) 94 (1972), 257-268.

70. L. Erbe, Oscillation criteria for second order nonlinear delay equations, Canad. Math. Bull. 16 (1973), 49-56.

71. L. Erbe and J. S. Muldowney, On the existence of oscillatory solutions to nonlinear differential equations, Ann. Mat. Pura Appl., to appear.

72. W. B. Fite, Concerning the zeros of the solutions of certain differential equations, Trans. Amer. Math. Soc. 19 (1918), 341-352.

73. K. Foster, Criteria for oscillation and growth of nonoscillatory solutions of forced differential equations of even order, J. Differential Equations 20 (1976), 115-132.

74. K. Foster, Oscillations of forced sublinear differential equations of even order, J. Math. Anal. Appl., to appear.

75. J. B. Garner, Oscillatory criteria for a general second order functional differential equation, SIAM J. Appl. Math. 29 (1975), 690-698.

76. H. E. Gollwitzer, On nonlinear oscillations for a second order delay equation, J. Math. Anal. Appl. 26 (1969), 385-389.

77. H. E. Gollwitzer, On the nonoscillatory behaviour of a second order linear equation, Quart. J. Math. Oxford Ser. (2) 21 (1970), 125-128.

78. H. E. Gollwitzer, Nonoscillation theorems for a nonlinear differential equation, Proc. Amer. Math. Soc. 26 (1970), 78-84.

79. H. E. Gollwitzer, Growth estimates for nonoscillatory solutions of a nonlinear differential equation, Časopis Pešt. Mat. 96 (1971), 119-125.

80. J. R. Graef, On the Generalized Liénard equation with negative damping, J. Differential Equations 12 (1972), 34-62.

81. J. R. Graef, Oscillation, nonoscillation, and growth of solutions of nonlinear functional differential equations of arbitrary order, J. Math. Anal. Appl., to appear.

82. J. R. Graef, Some nonoscillation criteria for higher order nonlinear differential equations, Pacific J. Math., to appear.

83. J. R. Graef, A comparison and oscillation result for second order nonlinear differential equations, to appear.

84. J. R. Graef and P. W. Spikes, A nonoscillation result for second order ordinary differential equations, Rend. Accad. Sci. Fis. Mat. Napoli (4) 41 (1974), 3-12.

85. J. R. Graef and P. W. Spikes, Asymptotic behaviour of solutions of a second order nonlinear differential equation, J. Differential Equations 17 (1975), 461-476.

86. J. R. Graef and P. W. Spikes, Nonoscillation theorems for forced second order nonlinear differential equations, Atti Accad. Naz. Lincei Rend. Cl. Sci. Fis. Mat. Natur., to appear.

87. J. R. Graef and P. W. Spikes, Asymptotic properties of solutions of functional differential equations of arbitrary order, J. Math. Anal. Appl., to appear.

88. M. K. Grammatikopoulos, Oscillatory and asymptotic behaviour of differential equations with deviating arguments, Hiroshima Math. J. 6 (1976), 31-53.

89. M. K. Grammatikopoulos, Y. G. Sficas, and V. A. Staikos, Oscillatory properties of strongly superlinear differential equations with deviating arguments, Univ. Ioannina, Tech. Report No. 36, 1975.

90. M. K. Grammatikopoulos, Y. G. Sficas, and V. A. Staikos, Asymptotic and oscillatory criteria for retarded differential equations, Univ. Ioannina, Tech. Report No. 37, 1975.

91. M. K. Grammatikopoulos, Y. G. Sficas, and V. A. Staikos, On the types of nonoscillatory solutions of differential equations with deviating arguments, University of Ioannina, Tech. Report No. 46, 1975.

92. G. W. Grefsrud, Existence and oscillation of solutions of certain functional differential equations, Ph.D. Thesis, Montana State Univ., Bozeman, 1971.

93. G. W. Grefsrud, Oscillatory properties of solutions of certain n-th order functional differential equations, Pacific J. Math. 60 (1975), 83-93.

94. R. Grimmer, On nonoscillatory solutions of a nonlinear differential equation, Proc. Amer. Math. Soc. 34 (1972), 118-120.

95. R. Grimmer, Oscillation criteria and growth of non-oscillatory solutions of even order ordinary and delay-differential equations, Trans. Amer. Math. Soc. 198 (1974), 215-228.

96. R. Grimmer and W. T. Patula, Nonoscillatory solutions of forced second order linear equations, to appear.

97. G. B. Gustafson, Oscillation criteria for $y''+p(t)f(y,y') = 0$ with f homogeneous of degree one, Canadian J. Math. 25 (1973), 323-337.

98. G. B. Gustafson, Bounded oscillations of linear and non-linear delay differential equations of even order, J. Math. Anal. Appl. 46 (1974), 175-189.

99. M. E. Hammett, Oscillation and nonoscillation theorems for nonhomogeneous linear differential equations of second order, Ph.D. Dissertation, Auburn University, Auburn, 1967.

100. M. E. Hammett, Nonoscillation properties of a nonlinear differential equation, Proc. Amer. Math. Soc. 30 (1971), 92-96.

101. J. W. Heidel, Qualitative behaviour of solutions of a third order nonlinear differential equation, Pacific J. Math. 27 (1968), 507-526.

102. J. W. Heidel, A nonoscillation theorem for a nonlinear second order differential equation, Proc. Amer. Math. Soc. 22 (1969), 485-488.

103. J. W. Heidel, The existence of oscillatory solutions for a nonlinear odd order differential equation, Czechoslvak Math. J. 20 (1970), 93-97.

104. J. W. Heidel, Rate of growth of nonoscillatory solutions for the differential equation $y''+q(t)|y|^{\gamma}\operatorname{sgn} y = 0$, $0<\gamma<1$, Quart. J. Appl. Math. 28 (1971), 601-606.

105. J. W. Heidel and D. B. Hinton, Existence of oscillatory solutions for a nonlinear differential equation, SIAM J. Math. Anal. 3 (1972), 344-351.

106. J. W. Heidel and I. T. Kiguradge, Oscillatory solutions for a generalized sublinear second order differential equation, Proc. Amer. Math. Soc. 38 (1973), 80-82.

107. Y. Hino, On oscillation of the solution of second order functional differential equations, Funkcial. Ekvac. 17 (1974), 95-105.

108. D. Hinton, An oscillation criterion for solutions of $(ry')'+qy^{\gamma} = 0$, Michigan Math. J. 16 (1969), 349-352.

109. J. W. Hooker, Existence and oscillation theorems for a class of non-linear second order differential equations, J. Differential Equations 5 (1969), 283-306.

110. H. C. Howard, Oscillation and nonoscillation criteria for $y''(x)+f(y(x))\cdot p(x) = 0$, Pacific J. Math. 12 (1962), 243-251.

111. H. C. Howard, Oscillation criteria for even order differential equations, Ann. Mat. Pura Appl. (4) 66 (1964), 221-231.

112. H. C. Howard, Oscillation and nonoscillation criteria for nonhomogeneous differential equations, to appear.

113. H. C. Howard and T. L. Hayden, Oscillation of differential equations in Banach spaces, Ann. Mat. Pura Appl. (4) 85 (1970), 383-394.

114. D. V. Izumova, On conditions for oscillation and nonoscillation for solutions of a nonlinear differential equation of the second order, Differencial'nye Uravnenija 12 (1966), 1572-1586.

115. D. V. Izumova and I. T. Kiguradge, Some remarks on the solutions of the equation $u'' + a(t)f(u) = 0$, Differencial'nye Uravnenija 4 (1968), 589-605.

116. M. Jasny, On the existence of an oscillatory solution of the nonlinear differential equation of second order $y'' + f(x)y^{2n-1} = 0$, $f(x)>0$, Časopis Pešt. Mat. 85 (1960), 78-83.

117. J. Jones, Jr., On the extension of a theorem of Atkinson, Quart. J. Math. Oxford Ser. (2) 7 (1956), 306-309.

118. J. Jones, Jr., On a nonlinear second order differential equation, Proc. Amer. Math. Soc. 9 (1958), 586-588.

119. I. V. Kamenev, On oscillatory solutions of nonlinear equations of second order, Trans. Moscow Electrical - Machine Design Institute 5 (1969/70), 125-136.

120. I. V. Kamenev, On some criteria of oscillation of solutions of ordinary differential equations of second order, Mat. Zametki 8 (1970), 773-776.

121. I. V. Kamenev, Oscillation of solutions of second-order nonlinear equations with sign-variable coefficients, Differencial'nye Uravnenija 6 (1970), 1718-1721.

122. I. V. Kamenev, Oscillation of solutions of a high-order nonlinear equation, Differencial'nye Uravnenija 7 (1971), 927-929.

123. I. V. Kamenev, A sufficient condition for oscillation of solutions of a high-order differential equation, Mat. Zametki 9 (1971), 421-423.

124. I. V. Kamenev, On some specific nonlinear oscillation theorems, Mat. Zametki 10 (1971), 129-134.

125. I. V. Kamenev, On integral criteria for nonoscillation, Mat. Zametki 13 (1973), 51-54.

126. G. A. Kamenskii, On the solutions of a linear homogeneous second order differential equation of the unstable type with retarded argument, Trudy Sem. Teor, Differents. Uravn. Otklon. Argum. 2 (1963), 82-93.

127. A. G. Kartsatos, Some theorems on oscillations of certain nonlinear second order ordinary differential equations, Archiv Math. 18 (1967), 425-429.

128. A. G. Kartsatos, On the relation between boundedness and oscillation of differential equations of second order, Canad. Math. Bull. 10 (1967), 675-679.

129. A. G. Kartsatos, Properties of bounded solutions of nonlinear equations of second order, Proc. Amer. Math. Soc. 19 (1968), 1057-1059.

130. A. G. Kartsatos, On oscillations of nonlinear equations of second order, J. Math. Anal. Appl. 24 (1968), 663-668.

131. A. G. Kartsatos, On oscillation and boundedness of solutions of second order nonlinear equations, Boll. Un. Mat. Ital. (4) 4 (1968), 357-361.

132. A. G. Kartsatos, Criteria for oscillation of solutions of differential equations of arbitrary order, Proc. Japan Acad. 44 (1968), 599-602.

133. A. G. Kartsatos, On oscillations of even order nonlinear differential equations, J. Differential Equations 6 (1969), 232-237.

134. A. G. Kartsatos, Contributions to the research of the oscillation and the asymptotic behaviour of solutions of ordinary differential equations, Bull. Soc. Math. Grèce 10 (1969), 1-48. (Greek).

135. A. G. Kartsatos, Oscillation properties of solutions of even order differential equations, Bull. Fac. Sci. Ibaraki Univ. Ser. A (1969), 9-14.

136. A. G. Kartsatos, On the maintenance of oscillations of nth order equations under the effect of a small forcing term, J. Differential Equations, 10 (1971), 355-363.

137. A. G. Kartsatos, Oscillation of nonlinear systems of matrix differential equations, Proc. Amer. Math. Soc. 30 (1971), 97-101.

138. A. G. Kartsatos, On the maintenance of oscillations under the effect of a periodic forcing term, Proc. Amer. Math. Soc. 34 (1972), 377-383.

139. A. G. Kartsatos, On positive solutions of perturbed nonlinear differential equations, J. Math. Anal. Appl. 47 (1974), 58-68.

140. A. G. Kartsatos, On nth order differential inequalities, J. Math. Anal. Appl. 52 (1975), 1-9.

141. A. G. Kartsatos, Oscillation and existence of unique positive solutions for nonlinear nth order equations with forcing term, Hiroshima Math. J. 6 (1976), 1-6.

142. A. G. Kartsatos, Analysis of the effect of certain forcings on the nonoscillatory solutions of even order equations, to appear.

143. A. G. Kartsatos, Oscillation of nth order equations with perturbations, J. Math. Anal. Appl. to appear.

144. A. G. Kartsatos, Oscillation and nonoscillation for perturbed differential equations, to appear.

145. A. G. Kartsatos, Nth order oscillations with middle terms of order n-2, to appear.

146. A. G. Kartsatos, Analysis of the effect of certain forcings on the nonoscillatory solutions of even order equations, to appear.

147. A. G. Kartsatos, On the behaviour near infinity of oscillatory solutions of functional n-th order equations, to appear.

148. A. G. Kartsatos and M. N. Manougian, Perturbations causing oscillations of functional-differential equations, Proc. Amer. Math. Soc. 43 (1974), 111-117.

149. A. G. Kartsatos and M. N. Manougian, Further results on oscillation of functional differential equations, J. Math. Anal. Appl. 53 (1976), 28-37.

150. A. G. Kartsatos and H. Onose, On the maintenance of oscillations under the effect of a small nonlinear damping, Bull. Faculty Sci. Ibaraki Univ. Ser. A (1972), 1-11.

151. A. G. Kartsatos and H. Onose, Remarks on oscillation of second order differential equations, Bull. Faculty Sci. Ibaraki Univ. Ser. A (973), 23-31.

152. A. G. Kartsatos and J. Toro, Oscillation and asymptotic behaviour of forced nonlinear equations, to appear.

153. A. G. Katranov, On zeros of oscillatory solutions of the equation $x'' + a(t)f(x) = 0$, Differencial'nye Uravnenija 7 (1971), 930-933.

154. A. G. Katranov, On the question of asymptotic behaviour of oscillatory solutions of nonlinear differential equations of second order, Differencial'nye Uravnenija 8 (1972), 785-789.

155. A. G. Katranov, On asymptotic behaviour of oscillatory solutions of the equation $x''+f(t,x)g(\dot{x}) = 0$, Differencial'nye Uravnenija 8 (1972), 1111-1115.

156. M. S. Kenner, On the solutions of certain linear nonhomogeneous second-order differential equations, Applicable Anal. 1 (1971), 57-63.

157. I. T. Kiguradge, On oscillatory solutions of some ordinary differential equations, Soviet Math. Dokl. 144 (1962), 33-36.

158. I. T. Kiguradge, On conditions for oscillation of solutions of the equation $u'' + a(t)|u|^n \text{sgn} u = 0$, Časopis Pešt. Mat. 87 (1962), 492-495.

159. I. T. Kiguradge, On oscillatory solutions of the equation $d^m u/dt^m + a(t)|u|^n \text{sign} u = 0$, Mat. Sbornik 65 (107)(1964), 172-187.

160. I. T. Kiguradge, On the question of oscillation of solutions of nonlinear differential equations, Differencial'nye Uravnenija 1 (1965), 995-1006.

161. I. T. Kiguradge, A note on the oscillation of solutions of the equation $u'' + a(t)|u|^n \text{ sgn} u = 0$, Časopis Pešt. Mat. 92 (1967), 343-350.

162. I. T. Kiguradge, The oscillatoriness of solutions of nonlinear ordinary differential equations, Proc. Fifth Inter. Conf. Nonlinear Oscillations, Vol. 1, Kiev, 1970, 293-298.

163. A. Kneser, Untersuchungen uber die reellen Nullstellen der Integrale linearer Differentialgleichungen, Math. Annalen 42 (1893), 409-435.

164. V. Komkov, On boundedness and oscillation of the differential equation $x'' + A(t)g(x) = f(t)$ in R^n, SIAM J. Appl. Math. 22 (1972), 561-568.

165. V. Komkov, On the location of zeros of second order differential equations, Proc. Amer. Math. Soc. 35 (1972), 217-222.

166. V. Komkov, A technique for the detection of oscillation of second order ordinary differential equations, Pacific J. Math. 42 (1972), 105-115.

167. V. Komkov, A generalization of Leighton's variational theorem, Applicable Anal. 1 (1972), 1-7.

168. V. Komkov, Asymptotic behaviour of nonlinear inhomogeneous differential equations via non-standard analysis, Part II. Ann. Polon. Math. 30 (1974), 93-106.

169. V. Komkov and C. Waid, Asymptotic behaviour of nonlinear inhomogeneous equations via non-standard analysis, Part I. Ann. Polon. Math. 28 (1973), 67-86.

170. V. A. Kondrat'ev, Oscillatory properties of solutions of the equation $y^{(n)} + p(x)y = 0$, Trudy Moskov Mat. Obšč. 10 (1961), 419-436.

171. R. G. Koplatadge, On the existence of oscillatory solutions of second order nonlinear differential equations with retarded argument, Soviet Math. Dokl. 210 (1973), 260-262.

172. R. G. Koplatadge, On oscillatory solutions of second order delay differential inequalities, J. Math. Anal. Appl. 42 (1973), 148-157.

173. R. G. Koplatadge, Remarks on the oscillation of solutions of higher order differential inequalities and equations with retarded argument, Differencial'nye Uravneniya 10 (1974), 1400-1405.

174. G. C. T. Kung, Oscillation and nonoscillation of differential equations with time lag, SIAM J. Appl. Math. 21 (1971), 207-213.

175. J. Kurzweil, A note on oscillatory solution of equation $y'' + f(x)y^{2n-1} = 0$, Časopis Pešt. Mat. 85 (1960), 357-358.

176. T. Kusano, Oscillatory behavior of solutions of higher-order retarded differential equations, Proc. C. Caratheodory Symp., Math. Soc. Grèce, Athens, 1973, 370-389.

177. T. Kusano and M. Naito, Nonlinear oscillation of second order differential equations with retarded argument, Ann. Mat. Pura Appl. (4) 106 (1975), 171-185.

178. T. Kusano and M. Naito, Nonlinear oscillation of fourth order differential equations, Canad. J. Math., to appear.

179. T. Kusano and H. Onose, Oscillation of solutions of nonlinear differential delay equations of arbitrary order, Hiroshima Math. J. 2 (1972), 1-13.

180. T. Kusano and H. Onose, Oscillation theorems for delay equations of arbitrary order, Hiroshima Math. J. 2 (1972), 263-270.

181. T. Kusano and H. Onose, Nonlinear oscillation of a sublinear delay equation of arbitrary order, Proc. Amer. Math. Soc. 40 (1973), 219-224.

182. T. Kusano and H. Onose, An oscillation theorem for differential equations with deviating argument, Proc. Japan Acad. 50 (1974), 809-811.

183. T. Kusano and H. Onose, Oscillations of functional differential equations with retarded argument, J. Differential Equations 15 (1974), 269-277.

184. T. Kusano and H. Onose, Oscillatory and asymptotic behavior of sublinear retarded differential equations, Hiroshima Math. J. 4 (1974), 343-355.

185. T. Kusano and H. Onose, Nonoscillatory solutions of differential equations with retarded arguments, Bull. Fac. Sci. Ibaraki Univ. Ser. A (1975), 1-11.

186. T. Kusano and H. Onose, Asymptotic behaviour of nonoscillatory solutions of second order functional differential equations, Bull. Austral. Math. Soc. 13 (1975), 291-299.

187. T. Kusano and H. Onose, Remarks on the oscillatory behaviour of solutions of functional differential equations with deviating argument, Hiroshima Math. J. 6 (1976), 183-189.

188. T. Kusano and H. Onose, Nonoscillation theorems for differential equations with deviating argument, Pacific J. Math. 63 (1976), 185-192.

189. T. Kusano and H. Onose, Asymptotic behavior of non-oscillatory solutions of functional differential equations of arbitrary order, J. London Math. Soc., to appear.

190. T. Kusano and H. Onose, Asymptotic behavior of non-oscillatory solutions of nonlinear differential equations with forcing term, Ann. Mat. Pura Appl., to appear.

191. T. Kusano, H. Onose, and H. Tobe, On the oscillation of second order nonlinear ordinary differential equations, Hiroshima Math. J. 4 (1974), 491-499.

192. G. Ladas, On oscillation and boundedness of solutions of nonlinear differential equations, Bull. Soc. Math. Grèce 10 (1969), 48-54.

193. G. Ladas, Oscillation and asymptotic behaviour of solutions of differential equations with retarded argument, J. Differential Equations 10 (1971), 281-290.

194. G. Ladas, G. Ladde, and J. S. Papadakis, Oscillations of functional-differential equations generated by delays, J. Differential Equations 12 (1972), 385-395.

195. G. Ladas and V. Lakshmikantham, Oscillations caused by retarded actions, Applicable Anal. 4 (1974), 9-15.

196. G. Ladas, V. Lakshmikantham, and J. S. Papadakis, Oscillations of higher-order retarded differential equations generated by the retarded argument, in Delay and Functional Differential Equations and Their Applications, Academic, New York, 1972, 219-231.

197. G. S. Ladde, Oscillations of nonlinear functional differential equations generated by retarded actions, in Delay and Functional Differential Equations and Their Applications, Academic, New York, 1972, 355-365.

198. B. S. Lalli and R. P. Jahagirdar, Comparison theorems of Levin type, J. Math. Anal. Appl. 49 (1975), 705-709.

199. A. Lasota, On convergence to zero of oscillating integrals of an ordinary differential equation of the second order, Zeszyty Nauk Uniw. Jagiell. Prace Mat. 6 (1961), 27-33.

200. G. G. Legatos and A. G. Kartsatos, Further results on oscillation of solutions of second order equations, Math. Japon. 14 (1968), 67-73.

201. W. Leighton, The detection of the oscillation of solutions of a second order linear differential equation, Duke Math. J. 17 (1950), 57-62.

202. W. Leighton, Comparison theorem for linear differential equations of second order, Proc. Amer. Math. Soc. 13 (1962), 603-610.

203. I. Ličko and M. Švec, Le caractere oscillatorie des solutions de l'equation $y^{(n)} + f(x)y^{\alpha} = 0$, $n>1$, Czechoslovak Math. J. 13 (1963), 481-489.

204. G. E. Liossatos, Some oscillation theorems for second order nonlinear differential equations with functional argument, Bull. Soc. Math. Grèce 11 (1970), 61-65.

205. G. E. Liossatos, Qualitative behaviour of solutions on nonlinear differential equations of order n, Doct. Dissertation, Univ. of Athens, Athens, Greece, 1973.

206. Tsai-Sheng Liu, Oscillation of even order differential equations with deviating arguments, Pacific J. Math. 61 (1975), 493-502.

207. Tsai-Sheng Liu, Classification of solutions of higher order differential equations and inequalities with deviating arguments, J. Differential Equations 21 (1976), 417-430.

208. Tsai-Sheng Liu, Oscillation of solutions of higher order functional differential equations and damped differential equations, to appear.

209. Stig-Olof Londen, Some nonoscillation theorems for a second order nonlinear differential equation, SIAM J. Math. Anal. 4 (1973), 460-465.

210. D. L. Lovelady, On the oscillatory behaviour of bounded solutions of higher order differential equations, J. Differential Equations 19 (1975), 167-175.

211. D. L. Lovelady, Asymptotic analysis of a second order nonlinear functional differential equation, Funkc. Ekvacioj 18 (1975), 15-22.

212. D. L. Lovelady, Nonoscillation in linear second order differential equations, Hiroshima Math. J. 5 (1975), 135-139.

213. D. L. Lovelady, Oscillation and a class of linear delay equations, to appear.

214. D. L. Lovelady, Oscillation and a class of forced even order differential equations, to appear.

215. D. L. Lovelady, Positive bounded solutions for a class of linear delay differential equations, to appear.

216. M. Luczynski, On the convergence to zero of oscillating solutions of an ordinary differential equation of order n, Zeszyty Nauk Uniw. Jagiell. Prace Mat. 7 (1962), 17-20.

217. J. W. Macki and J. S. W. Wong, Oscillation of solutions to second-order nonlinear differential equations, Pacific J. Math. 24 (1968), 111-117.

218. W. Mahfoud, A noncontinuation criterion for an n-th order equation with a retarded argument, This Volume, Chapter 11.

219. W. E. Mahfoud, Noncontinuation and uniqueness of solutions of the delay equation $x^{(n)}(t) + a(t)\phi(x(t))f(x(q(t))) = 0$, J. Math. Anal. Appl., to appear.

220. W. E. Mahfoud, Oscillation and asymptotic behavior of solutions of nth order nonlinear delay differential equations, J. Differential Equations, to appear.

221. W. E. Mahfoud, Remarks on some oscillation theorems for nth order differential equations with a retarded argument, to appear.

222. P. Marusiak, Note on the Ladas' paper on oscillation and asymptotic behaviour of solutions of differential equations with retarded argument, J. Differential Equations 13 (1973), 150-156.

223. P. Marusiak, Oscillation of solutions of the delay differential equation $y^{(2n)}(t) + \sum_{i=1}^{m} P_i(t)f_j(y[h_i(t)]) = 0$, $n>1$, Časopis Pešt. Mat. 99 (1974), 131-141.

224. J. G. Mikusinski, On Fite's oscillation theorems, Colloq. Math. 2 (1949), 34-39.

225. W. E. Milne, A theorem on oscillation, Bull. Amer. Math. Soc. 28 (1922), 102-104.

226. R. A. Moore, The behaviour of solutions of a linear differential equation of second order, Pacific J. Math. 5 (1955), 125-145.

227. R. A. Moore and Z. Nehari, Non-oscillation theorems for a class of nonlinear differential equations, Trans. Amer. Math. Soc. 93 (1959), 30-52.

228. M. Naito, Oscillation theorems for a damped nonlinear differential equation, Proc. Japan Acad. 50 (1974), 104-108.

229. M. Naito, Oscillations of differential inequalities with retarded arguments, Hiroshima Math. J. 5 (1975), 187-192.

230. M. Naito, Oscillation criteria for a second order differential equation with a damping term, to appear.

231. Z. Nehari, On a class of nonlinear second-order differential equations, Trans. Amer. Math. Soc. 95 (1960), 101-123.

232. Z. Nehari, Oscillation theorems for systems of linear equations, Trans. Amer. Math. Soc. 139 (1969), 339-347.

233. Z. Nehari, A nonlinear oscillation problem, J. Differential Equations 5 (1969), 452-460.

234. Z. Nehari, Oscillation and boundedness criteria for a class of nonlinear differential systems, J. Differential Equations 11 (1972), 1-9.

235. Z. Nehari, A nonlinear oscillation theorem, Duke Math. J. 42 (1975), 183-189.

236. O. N. Odarič and V. N. Ševelo, Certain problems of the asymptotic behaviour of solutions of nonlinear differential equations with retarded argument, Differencial'nye Uravnenija 9 (1973), 637-646.

237. S. N. Olekhnik, Oscillatory character of solutions of a system of ordinary differential equations of second order, Differencial'nye Uravnenija 9 (1973), 2146-2151.

238. H. Onose, Oscillatory property of certain nonlinear ordinary differential equations, Proc. Japan Acad. 44 (1968), 110-113.

239. H. Onose, Oscillatory property of certain nonlinear ordinary differential equations, II, Proc. Japan Acad. 44 (1968), 876-878.

240. H. Onose, Oscillatory property of ordinary differential equations of arbitrary order, J. Differential Equations 7 (1970), 454-458.

241. H. Onose, Oscillatory property of certain nonlinear ordinary differential equations, SIAM J. Appl. Math. 18 (1970), 715-719.

242. H. Onose, Oscillation theorems for nonlinear second order differential equations, Proc. Amer. Math. Soc. 26 (1970), 461-464.

243. H. Onose, Oscillatory properties of solutions of even order differential equations, Pacific J. Math. 38 (1971), 747-757.

244. H. Onose, Some oscillation criteria for nth order nonlinear delay-differential equations, Hiroshima Math. J. 1 (1971), 171-176.

245. H. Onose, On oscillations for solutions of nth order differential equations, Proc. Amer. Math. Soc. 33 (1972), 495-500.

246. H. Onose, On oscillations of nonlinear second-order equations, J. Math. Anal. Appl. 39 (1972), 122-124.

247. H. Onose, Oscillation and asymptotic behaviour of solutions of retarded differential equations of arbitrary order, Hiroshima Math. J. 3 (1973), 333-360.

248. H. Onose, A comparison theorem and the forced oscillation, Bull. Austral. Math. Soc. 13 (1975), 13-19.

249. H. Onose, Oscillation criteria for second order nonlinear differential equations, Proc. Amer. Math. Soc. 51 (1975), 67-73.

250. H. Onose, Oscillation and nonoscillation of delay differential equations, Ann. Mat. Pura Appl., to appear.

251. Z. Opial, Sur les intégrales oscillantes de l'équation différentielle u" + f(t)u = 0, Ann. Polon. Math. 4 (1958), 308-313.

252. Z. Opial, Sur une critère d'oscillation de l'équation différentielle (Q(t)x')' + f(t)x = 0, Ann. Polon. Math. 6 (1959), 99-104.

253. W. F. Osgood, On a theorem of oscillation, Bull. Amer. Math. Soc. 25 (1919), 216-221.

254. R. V. Petropavlovskaya, On the oscillatory aspect of solutions of the differential equation u" = f(u,u',t), Dokl. Akad. Nauk SSSR 108 (1956), 389-391.

255. C. M. Petty and W. E. Johnson, Properties of solutions of u" + c(t)f(u)h(u') = 0 with explicit initial conditions, SIAM J. Math. Anal. 4 (1973), 269-282.

256. L. Pintér, Oszillationssätze fur einen Typ von nichtlinearen Differentialgleichungen sweiter ordnung, Magyar Tud. Akad. Mat. Kutato Int. Kozl. 6 (1961), 333-350.

257. S. M. Rankin, Oscillation of a forced second order nonlinear differential equation, Proc. Amer. Math. Soc., to appear.

258. G. H. Ryder and D. V. V. Wend, Oscillation of solutions of certain ordinary differential equations of nth order, Proc. Amer. Math. Soc. 21 (1970), 463-469.

259. V. M. Sakhare, Asymptotic and oscillation theorems for a second order differential equation with delay, Doct. Dissertation, Univ. of Tennessee, Knosville, 1973.

260. V. N. Ševelo, Some questions of the theory of oscillation of solutions of nonlinear nonautonomous differential equations, Proc. Fourth Inter. Conf. Nonlinear Oscillations, Prague, 1968, 251-256.

261. V. N. Ševelo, On the influence of retarded arguments on the oscillation of solutions of differential equations of even order, Proc. Fifth Inter. Conf. Nonlinear Oscillation, Kiev, 1970, 553-557.

262. V. N. Ševelo and O. N. Odarič, Some questions in the theory of oscillation (nonoscillation) of solutions of differential equations of second order, Ukrain. Mat. Ž. 23 (1971), 508-516.

263. V. N. Ševelo and V. N. Vareh, On the oscillation of solutions of higher order linear differential equations with retarded argument, Ukrain. Mat. Ž. 24 (1972), 513-520.

264. V. N. Ševelo and V. N. Vareh, On some properties of solutions of differential equations with delay, Ukrain. Mat. Ž. 24 (1972), 807-813.

265. Y. G. Sficas, On oscillation and asymptotic behaviour of certain class of differential equations with retarded argument, Utilitas Math. 3 (1973), 239-249.

266. Y. G. Sficas, An oscillation criterion for second order delay differential equations, Bull. Soc. Math. Grèce 12 (1974), 41-44.

267. Y. G. Sficas, Retarded actions on oscillations, Univ. of Ioannina, Tech. Report No. 15, 1974.

268. Y. G. Sficas, The effect of the delay on the oscillatory and asymptotic behaviour of nth order retarded differential equations, J. Math. Anal. Appl. 49 (1975), 748-757.

269. Y. G. Sficas, On the oscillatory and asymptotic behaviour of damped differential equations with retarded argument, Univ. of Ioannina, Tech. Report No. 41, 1975.

270. Y. G. Sficas and V. A. Staikos, Oscillations of retarded differential equations, Proc. Camb. Phil. Soc. 75 (1974), 95-101.

271. Y. G. Sficas and V. A. Staikos, Oscillations of differential equations with retardations, Hiroshima Math. J. 4 (1974), 1-8.

272. Y. G. Sficas and V. A. Staikos, The effect of retarded actions on nonlinear oscillations, Proc. Amer. Math. Soc. 46 (1974), 259-264.

273. Y. G. Sficas and V. A. Staikos, Oscillations of differential equations with deviating arguments, Funkcial. Ekvac., to appear.

274. K. D. Shere, Nonoscillation of second order linear differential equations with retarded argument, J. Math. Anal. Appl. 41 (1973), 293-299.

275. W. E. Shreve, Oscillation in first order nonlinear retarded argument differential equations, Proc. Amer. Math. Soc. 40 (1973), 565-568.

276. B. Singh, A necessary and sufficient condition for the oscillation of an even order nonlinear delay differential equation, Can. J. Math 25 (1973), 1078-1089.

277. B. Singh, Oscillation and nonoscillation of even order nonlinear delay-differential equations, Quart. Appl. Math. 31 (1973), 343-349.

278. B. Singh, Asymptotic nature of nonoscillatory solutions of nth order retarded differential equations, SIAM J. Math. Anal. 6 (1975), 784-795.

279. B. Singh, Comparison theorems for even order differential delay equations, Bull. Cal. Math. Soc. 67 (1975), 23-28.

280. B. Singh, Impact of delays on oscillation in general functional equations, Hiroshima Math. J. 5 (1975), 351-361.

281. B. Singh, Damp nonoscillation in third order retarded equations, Funkcial. Ekvac. 18 (1975), 5-14.

282. B. Singh, Forced oscillations in general ordinary differential equations with deviating arguments, Hiroshima Math. J. 6 (1976), 7-14.

283. B. Singh and R. S. Dahiya, On oscillation of second-order retarded equations, J. Math. Anal. Appl. 47 (1974), 504-512.

284. B. Singh and R. S. Dahiya, Nonoscillation of third order retarded equations, Bull. Austral. Math. Soc. 10 (1974), 9-14.

285. P. W. Spikes, Behaviour of the solutions of the differential equation $y'' + qy^p = r$, Applicable Anal. 4 (1974), 253-264.

286. V. A. Staikos, Oscillatory property of certain delay-differential equations, Bull. Soc. Math. Grèce 11 (1970), 1-5.

287. V. A. Staikos and A. G. Petsoulas, Some oscillation criteria for second order nonlinear delay differential equations, J. Math. Anal. Appl. 30 (1970), 695-701.

288. V. A. Staikos and Ch. G. Philos, On the asymptotic behavior of nonoscillatory solutions of differential equations with deviating arguments, Univ. of Ioannina, Tech. Report No. 45, 1975.

289. V. A. Staikos and Ch. G. Philos, Some oscillation and asymptotic properties for linear differential equations, Univ. of Ioannina, Tech. Report No. 54, 1976.

290. V. A. Staikos and Ch. G. Philos, Asymptotic properties of nonoscillatory solutions of differential equations with deviating arguments, Univ. of Ioannina, Tech. Report No. 55, 1976.

291. V. A. Staikos and Y. G. Sficas, Oscillatory and asymptotic behavior of functional differential equations, J. Differential Equations 12 (1972), 426-437.

292. V. A. Staikos and Y. G. Sficas, Criteria for asymptotic and oscillatory character of functional differential equations of arbitrary order, Boll. Un. Mat. Ital. 6 (1972), 185-192.

293. V. A. Staikos and Y. G. Sficas, Some results on oscillatory and asymptotic behavior of differential equations with deviating arguments, Proc. Caratheodory Symp., Math. Soc. Greece, Athens, 1973, 546-553.

294. V. A. Staikos and Y. G. Sficas, Oscillations for forced second order nonlinear differential equations, Atti Accad. Naz. Lincei Rend. Cl. Sci. Fis. Mat. Natur. (8) 55 (1973), 25-30.

295. V. A. Staikos and Y. G. Sficas, Oscillatory and asymptotic properties of differential equations with retarded arguments, Applicable Anal. 5 (1975), 141-148.

296. V. A. Staikos and Y. G. Sficas, Oscillatory and asymptotic characterization of the solutions of differential equations with deviating arguments, J. London Math. Soc. 10 (1975), 39-47.

297. V. A. Staikos and Y. G. Sficas, Forced oscillations for differential equations of arbitrary order, J. Differential Equations 17 (1975), 1-11.

298. V. A. Staikos and Y. G. Sficas, On the oscillation of bounded solutions of forced differential equations with deviating argument, Atti Accad. Naz. Lincei Rend. Cl. Sci. Fis. Mat. Natur. (8) 58 (1975), 318-322.

299. V. A. Staikos and I. P. Stavroulakis, Bounded oscillations under the effect of retardations for differential equations of arbitrary order, Univ. of Ioannina, Tech. Report No. 58, 1976.

300. I. P. Stavroulakis, Oscillatory and asymptotic properties of differential equations with deviating arguments, Univ. of Ioannina, Tech. Report No. 59, 1976.

301. E. Sturm, Sur les équations différentielles linéaires du second ordré, J. Math. Pures Appl. 1 (1836), 106-186.

302. M. Švec, Fixtpunktsatz und monotone Losungen der Differentialgleichungen $y^{(n)} + B(x,y,y',\dots,y^{(n-1)}) = 0$, Arch. Math. (Brno) 2 (1966), 43-55.

303. M. Švec, Monotone solutions of some differential equations, Colloq. Math. 18 (1967), 7-21.

304. M. Švec, Some oscillatory properties of second order non-linear differential equations, Ann. Mat. Pura Appl. (4) 77 (1967), 179-192.

305. M. Švec, Les propriétés asymptotiques des solutions d'une équation différentielle nonlinéaire d'ordre n, Czechoslovak Math. J. 17 (1967), 550-557.

306. C. A. Swanson, Comparison and oscillation theory of linear differential equations, Academic, New York, 1968.

307. S. C. Tefteller, Oscillation of second order nonhomogeneous differential equations, Ph.D. Dissertation, Univ. of Houston, Houston, 1972.

308. S. C. Tefteller, Oscillation of second order nonhomogeneous linear differential equations, SIAM J. Appl. Math., to appear.

309. S. C. Tefteller, Oscillation of n-th order nonhomogeneous linear differential equations, to appear.

310. R. D. Terry, Oscillatory properties of a fourth order delay differential equation, 2, Funkcial. Ekvac. 16 (1973), 213-224.

311. R. D. Terry, Oscillatory properties of a delay differential equation of even order, Pacific J. Math. 52 (1974), 269-282.

312. R. D. Terry, Some oscillation criteria for delay-differential equations of even order, SIAM J. Appl. Math. 28 (1975), 319-334.

313. R. D. Terry, Delay differential equations of odd order satisfying Property P_k, J. Austral. Math. Soc. 20 (1975), 451-467.

314. R. D. Terry and P. K. Wong, Oscillatory properties of a fourth order delay differential equation, Funkcial. Ekvac. 15 (1972), 209-221.

315. H. Teufel, Jr., Second order nonlinear oscillation-deviating arguments, Monatsh. Math. 75 (1971), 341-345.

316. H. Teufel, Jr., On the behavior of solutions of sublinear second order differential equations, Proc. Amer. Math. Soc. 32 (1972), 445-451.

317. H. Teufel, Jr., Second order almost linear functional equations-oscillation, Proc. Amer. Math. Soc. 35 (1972), 117-119.

318. H. Teufel, Jr., A note on second order differential inequalities and functional differential equations, Pacific J. Math. 41 (1972), 537-541.

319. H. Teufel, Jr., Forced second order nonlinear oscillation, J. Math. Anal. Appl. 40 (1972), 148-152.

320. H. Teufel, Jr., Remarks on damped nonlinear delay-oscillators, to appear.

321. E. C. Tomastik, Oscillation of a nonlinear second order differential equation, SIAM J. Appl. Math. 15 (1967), 1275-1277.

322. C. C. Travis, Oscillation theorems for second order differential equations with functional arguments, Proc. Amer. Math. Soc. 31 (1972), 199-202.

323. C. C. Travis, A note on second order nonlinear oscillations, Math. Japon. 18 (1973), 261-264.

324. W. F. Trench, Oscillation properties of perturbed dis-conjugate equations, Proc. Amer. Math. Soc. 52 (1975), 147-155.

325. E. True, Existence, comparison and oscillation results for some functional differential equations, Doctoral Dissertation, Montana State Unv., Bozeman, 1972.

326. E. True, A comparison theorem for certain functional differential equations, Proc. Amer. Math. Soc. 47 (1975), 127-132.

327. W. R. Utz, A note on second-order nonlinear differential equations, Proc. Amer. Math. Soc. 7 (1956), 1047-1048.

328. W. R. Utz, Properties of solutions of certain second order nonlinear differential equations, Proc. Amer. Math. Soc. 8 (1957), 1024-1028.

329. W. R. Utz, Properties of solutions of $u'' + g(t)u^{2n-1} = 0$, Monatsh. Math. 66 (1962), 55-60.

330. W. R. Utz, Properties of solutions of $u'' + g(t)u^{2n-1} = 0$, Monatsh. Math. 69 (1965), 353-361.

331. W. R. Utz, The behavior of solutions of the equations $x'' \pm xx'^2 \pm x^2 = 0$, Amer. Math. Monthly 74 (1967), 420-422.

332. W. R. Utz, Properties of solutions of certain nonlinear differential equations of the form $x'' + g(x)x'^2 + f(x) = 0$, Ann. Mat. Pura Appl. (4) 81 (1969), 61-68.

333. G. Villari, Sal comportamento asintotico degli integrali di una classe di equazioni differenziali non lineari, Rev. Mat. Univ. Parma 5 (1954), 83-98.

334. P. Waltman, Oscillation of solutions of nonlinear equations, SIAM Review 5 (1963), 128-130.

335. P. Waltman, Some properties of solutions of $u'' + a(t)f(u) = 0$, Monatsh. Math. 67 (1963), 50-54.

336. P. Waltman, On the asymptotic behavior of solutions of a nonlinear equation, Proc. Amer. Math. Soc. 15 (1964), 918-923.

337. P. Waltman, An oscillation criterion for a nonlinear second order equation, J. Math. Anal. Appl. 10 (1965), 439-441.

338. P. Waltman, Oscillation criteria for third order nonlinear differential equations, Pacific J. Math. 18 (1966), 385-389.

339. P. Waltman, A note on an oscillation criterion for an equation with a functional argument, Canad. Math. Bull. 11 (1968), 593-595.

340. D. Willett, Classification of second order linear differential equations with respect to oscillation, Advances in Math. 3 (1969), 594-693.

341. A. Wintner, A criterion of oscillatory stability, Quart. Appl. Math. 7 (1949), 115-117.

342. J. S. W. Wong, On second order nonlinear oscillation, Funkcial. Ekvac. 11 (1968), 207-234.

343. J. S. W. Wong, Second order oscillation with retarded arguments, in Ordinary Differential Equations 1971 NRL-MRC Conference, Academic, New York, 1972, 581-596.

344. J. S. W. Wong, On the generalized Emden-Fowler Equation, SIAM Review 17 (1975), 339-360.

345. J. S. W. Wong, Oscillation theorems for second order nonlinear differential equations, Bull. Inst. Math. Acad. Sinica 3 (1975), 283-309.

346. Pui-Kei Wong, Existence and asymptotic behavior of proper solutions of a class of second-order nonlinear differential equations, Pacific J. Math. 13 (1963), 737-760.

347. Pui-Kei Wong, On a class of nonlinear fourth order differential equations, Ann. Mat. Pura Appl. (4) 81 (1969), 331-346.

348. T. Yoshizawa, Oscillatory property of second order differential equations, Tohoku Math. J. 22 (1970), 619-634.

349. T. Yoshizawa, Oscillatory property for second order differential equations, In Ordinary Differential Equations 1971 NRC-MRC Conference, Academic, New York, 1972, 315-327.

350. M. Zlamal, Asymptotische Eigenschaften der Lösungen linearer Differentialgleichungen, Math. Nach. 10 (1953), 169-174.

Chapter 4

CONTRACTIVE MAPPINGS AND PERIODICALLY PERTURBED NON-CONSERVATIVE SYSTEMS

R. REISSIG

Institut für Mathematik
Ruhr-Universität Bochum
Federal Republic of Germany

INTRODUCTION

Let us consider the vector differential equation of Liénard type

$$x'' + Cx' + \text{grad } G(x) = e(t), \ x \in R^n \tag{1}$$

where $G \in C^2(R^n, R)$ with the Hessian matrix $H(x) = (h_{rs}(x))$, $e \in C^o(R, R^n)$, $e(t+2\pi) \equiv e(t)$, and $C = C^*$ is a constant real matrix. Let us look for conditions which ensure the existence of one and only one 2π-periodic solution. Especially we are interested in the construction of this solution by means of Picard's iteration.

There are some recent papers which are devoted to the forced vibration problem of the conservative system (1) corresponding to the special case $C = 0$ (see [1],[3],[5] - [8]). The most general condition was developed by A. C. Lazer [5]. Let the constant real matrices $P = P^*$ and $Q = Q^*$ be such that

$$Q \leq H(x) \leq P \text{ for all } x \in R^n \tag{2}$$

(where $Q \leq P$ means that $x^*(P - Q)x \geq 0$ for all $x \in R^n$) and let

$$m_r^2 < \lambda_r(Q) \leq \lambda_r(P) < (m_r+1)^2, \ 1 \leq r \leq n \tag{3}$$

(where $m_1, \ldots, m_n$ are non-negative integers). Assuming (2) and (3) Lazer [5] proved the uniqueness of the periodic solution whereas S. Ahmad [1] completed this result by solving the existence problem.

J. Mawhin [11] showed that the periodic solution can be constructed via successive approximations provided that the conditions (2), (3) are specialized as

$$m^2 I < qI \leq H(x) \leq pI < (m+1)^2 I \text{ for all } x \in R^n \tag{4}$$

(m a non-negative integer). The basic idea of Mawhin's proof is to generalize the periodic boundary value problem as an operator problem in a suitable Hilbert space. The linear differential operator of this problem is studied with the aid of spectral theory of self-adjoint unbounded operators. This procedure is no longer applicable when equation (1) contains a damping term ($C \neq 0$). But in the present paper we show that it can be replaced by a simpler argument based on Green functions which is, however, more efficient.

Let us start from a linear higher order differential operator with periodic boundary conditions in order to derive a corresponding mapping theorem in a Hilbert function space. This theorem will be applied to the Liénard equation (1) in case $n = 1$. As a complement of this study, we point out some further results concerning the periodic solutions of the generalized Liénard equation which contains an arbitrary nonlinear damping term and a less restrictive restoring term. Finally, we apply the results of case $n = 1$ in order to construct the 2π-periodic solution of the vector equation (1). This is possible when Lazer's condition is somewhat modified:

1. Assume that the symmetric matrices C, Q and P can be diagonalized by means of the same orthogonal transformation $y = Tx (T^* = T^{-1})$.
2. Replace condition (2) written in a symmetric form

$$-(P-Q)/2 \leq N - H(x) \leq (P-Q)/2 \text{ where } N = (P+Q)/2 \tag{5}$$

by a similar estimate which is, however, independent upon (5), see [9]:

$$(N - H(x))^2 \leq (P-Q)^2/4 \tag{6}$$

3. Admit a weaker version of the eigenvalue condition (3) and replace it (for some r) by

$$\lambda_r(Q) \leq \lambda_r(P) < 0 \tag{7}$$

<u>Remark.</u> Introducing y = Tx we obtain

$$y'' + (TCT^*)y' + \mathrm{grad}_y G(T^*y) = T(x'' + Cx' + \mathrm{grad}_x G(x))$$

since

$$\mathrm{grad}_y G(T^*y) = T\,\mathrm{grad}_x G(x)$$

Moreover, we have

$$\left(\frac{\partial^2}{\partial y_r \partial y_s} G(T^*y)\right) = T\left(\frac{\partial^2}{\partial x_r \partial x_s} G(x)\right) T^*$$

Consequently, the estimates (5) or (6) are invariant with respect to an orthogonal transformation.

A LINEAR PERIODIC BOUNDARY VALUE PROBLEM

Let us consider a linear differential operator with real constant coefficients which is defined on class C^k:

$$Lx \equiv x^{(k)} + a_1 x^{(k-1)} + \ldots + a_k x, \ k \geq 2 \tag{8}$$

The equation $Lx = e(t) \in C^o[0,2\pi]$ possesses exactly one solution satisfying the periodic boundary conditions

$$x^{(j)}(0) = x^{(j)}(2\pi), \ 0 \leq j \leq k-1 \tag{9}$$

provided that $p(im) \neq 0$ for all integers m where $p(\lambda)$ the characteristic polynomial. This solution can be represented by means of a Green function:

$$x(t) = \int_0^{2\pi} \gamma(t-s)e(s)ds$$

A similar representation is valid for the derivatives up to order k-1; the kernel is the corresponding derivative of the Green function which is continuous if the order doesn't exceed k-2, and which is piecewise continuous if the order is equal k-1:

$$\gamma^{(k-1)}(0+)-\gamma^{(k-1)}(2\pi-) = 1$$

Now the boundary value problem will be generalized as follows. Let H be the complex Hilbert space $L^2[0,2\pi]$ supplied with the inner product

$$\langle x,y\rangle = (2\pi)^{-1}\int_0^{2\pi} x(t)\overline{y}(t)\,dt$$

Let the subspace $D \subset H$ consist of all $x \in H$ possessing Lebesgue derivatives in H up to order k and satisfying the boundary conditions (9). Define the linear mapping

$$L : D \to H,\ x \to Lx$$

L is an injection since $Lx = 0$ implies $x(t) = 0$ for all $t \in [0,2\pi]$ by virtue of the assumption $p(im) \neq 0$ if m an integer. Let us show that L is a surjection, $L(D) = H$, and that the inverse mapping L^{-1} is compact.

Define the linear mappings

$$\Gamma_j : H \to H,\ y \to \int_0^{2\pi} \gamma^{(j)}(t-s)y(s)\,ds,\ 0 \leq j \leq k-1$$

which are bounded,

$$|(\Gamma_j y)(t)| \leq \rho_j ||y||,\ \rho_j = 2\pi||\gamma^{(j)}||,\ t \in [0,2\pi] \tag{10}$$

If $y \in C^0[0,2\pi]$ then $\Gamma_j y \in C^1$ $(0 \leq j \leq k-1)$ and

$$\begin{aligned} &\Gamma_j y|_0^t = \int_0^t (\Gamma_{j+1}y)(s)\,ds,\ \Gamma_j y|_0^{2\pi} = 0\ (0 \leq j \leq k-2) \\ &\Gamma_{k-1}y|_0^t = \int_0^t ((id_H - a_1\Gamma_{k-1} - \dots - a_k\Gamma_o)y)(s)\,ds,\ \Gamma_{k-1}y|_0^{2\pi} = 0 \end{aligned} \tag{11}$$

Since C^o dense in H, by virtue of (10) the equations (11) are also valid for an arbitrary $y \in H$. Consequently we have

$x = \Gamma y \in D$ (where the denotation $\Gamma = \Gamma_o$ is used)
$x^{(j)} = \Gamma_j y$ (absolutely continuous, $1 \leq j \leq k-1$)
$x^{(k)} = (id_H - a_1\Gamma_{k-1} - \dots - a_k\Gamma_o)y$ (derivative in Lebesgue's sense)

Hence

$$L(\Gamma y) = y \text{ for all } y \in H, \ L\Gamma = id_H$$
$$\Gamma(Lx) = x \text{ for all } x \in D, \ \Gamma L = id_D$$
$$\Gamma = L^{-1} \text{ (bounded linear operator on H)}$$

To see that the inverse operator L^{-1} is compact, let $\{y_n\}$ be a bounded sequence in H, $||y_n|| \leq R$ for all n. Let $\{x_n\}$ be its image, i.e. $x_n = \Gamma y_n$. Then, according to (10), the $x_n \in D$ are equibounded and equicontinuous,

$$|x_n(t)| \leq \rho_o R \text{ for all } n \in N, \text{ for all } t \in [0,2\pi]$$
$$|x_n(t)-x_n(s)| \leq \rho_1 R|t-s| \text{ for all n and for all}$$
$$(t,s) \in [0,2\pi]^2$$

There must be a uniformly convergent subsequence the limit of which is continuous and therefore an element of H. That is, $\Gamma = L^{-1}$ maps each bounded subset of H into a compact subset.

In order to calculate $||L^{-1}||$ let us consider an eigenvalue λ and a corresponding normed eigenfunction $\phi(t)$:

$$L^{-1}\phi = \lambda\phi$$

Since $\lambda \neq 0$, $\phi \in D$ and $\phi = \lambda L\phi$, or $(\lambda L - id_D)\phi = 0$. This is a homogeneous linear differential equation (in the classical sense) having only non-trivial solutions of the type exp(i m t), m an integer, when the periodic boundary conditions are prescribed. So, without loss of generality

$$\phi = \phi_m = e^{imt}, \ \lambda = \lambda_m = \frac{1}{p(im)}$$

Since the Hilbert space H is separable, and since the family $(\phi_m)_{m \in Z}$ is a complete orthonormal system each $y \in H$ can be represented as the sum of its Fourier series

$$y = \sum_{-\infty}^{+\infty} b_m\phi_m$$

Consequently,

$$L^{-1}y = \sum_{-\infty}^{+\infty} b_m L^{-1}\phi_m = \sum_{-\infty}^{+\infty} \lambda_m b_m \phi_m$$

$$|\lambda_{-p}| = |\lambda_p| \leq ||L^{-1}|| \leq \sup_{m\varepsilon N}|\lambda_m| \text{ for all } p \varepsilon N$$

and

$$||L^{-1}|| = \sup_{m\varepsilon N}|\lambda_m| \tag{12}$$

Consider, for example,

$$Lx \equiv x'' + cx' + \nu x \tag{13}$$

where c is an arbitrary real value, (i) $m^2 < \nu < (m+1)^2$ for a non-negative integer m, or (ii) $\nu < 0$. Then

$$|\lambda_p| = ((p^2-\nu)^2+c^2p^2)^{-1/2}$$

and

$$||L^{-1}|| \leq \alpha = \begin{cases} \max\left((\nu-m^2)^{-1},((m+1)^2-\nu)^{-1}\right), \text{ in case (i)} \\ |\nu|^{-1}, \text{ in case (ii)} \end{cases} \tag{14}$$

(norm of the operator when $c = 0$).

THE SCALAR LIENARD EQUATION

In order to solve the periodic boundary value problem of the scalar Liénard equation

$$x'' + cx' + g(x) = e(t) \equiv e(t+2\pi) \varepsilon C^0 \tag{15}$$

under the condition (i) $m^2 < q \leq g'(x) \leq p < (m+1)^2$, or, (ii) $q \leq g'(x) \leq p < 0$, we consider the generalized problem

$$Lx - Bx = y \varepsilon H \text{ (real Hilbert space } L^2[0,2\pi]) \tag{16}$$

where $x \varepsilon D$, Lx is given by (13), and

$$Bx = \nu x - g(x)$$

is a bounded nonlinear operator from H into H. This is due to the fact that the nonlinear restoring term g is continuous and linearly bounded: $|\nu x - g(x)| \leq |g(0)| + \beta|x|$ for $x \varepsilon R$ where $\beta = \max(|q-\nu|,|p-\nu|)$. Moreover, the nonlinear operator B is globally Lipschitzian:

$$||Bx_1 - Bx_2|| \leq \beta||x_1 - x_2|| \tag{17}$$

The solutions of equation (16) are the fixed points of the mapping

$$H \to H, \; x \to L^{-1}Bx + L^{-1}y$$

which is a contraction provided that $||L^{-1}B|| < 1$. This is ensured when $\alpha\beta < 1$, and it can be realized by a suitable choice of ν (see [11]): (i) $m^2+p < 2\nu < (m+1)^2+q$, or, (ii) $2\nu < q$. Then there is exactly one fixed point $x^* = L^{-1}Bx^* + L^{-1}y \in D$ which can be obtained by means of Picard's iteration. If $\{x_n\}_{n \in N}$ is a sequence of successive approximations then

$$\begin{aligned} |x^*(t)-x_{n+1}(t)| &\leq \rho_o\beta||x^* - x_n|| \\ |x^{*\prime}(t)-x'_{n+1}(t)| &\leq \rho_1\beta||x^* - x_n|| \end{aligned} \tag{18}$$

This means that the approximations $x_n(t)$ and their first derivatives converge uniformly on $[0,2\pi]$ in the usual sense.

Replacing $y \in H$ by a continuous 2π-periodic forcing term $e(t)$, we obtain the classical periodic solution of the initial equation (15).

J. Mawhin has mentioned that equation (16) has still a solution when the nonlinearity g fulfils a weaker condition:

$$g(x) \in C^o(R), \; q \leq \frac{g(x)}{x} \leq p \; (|x| \geq X) \tag{19}$$

where the bounds q and p are like before.

Defining

$$g(x) = g^*(x) + \rho(x), \; g^*(x) = \begin{cases} \frac{g(X \operatorname{sgn} x)}{X}, & |x| \leq X \\ g(x), & |x| \geq X \end{cases} \tag{20}$$

we can estimate

$$\begin{aligned} |\nu x - g(x)| &\leq |\nu x - g^*(x)| + |\rho(x)| \\ &\leq \beta|x| + P \end{aligned}$$

(with the same values ν and β as before). Therefore, the operator equation (16) written in the form

$$x = L^{-1}Bx + L^{-1}y \tag{21}$$

can be considered again. But this time we will apply Schauder's fixed point theorem with respect to a closed ball $\overline{B}_R \varepsilon H$ of sufficiently large radius R. For this reason we must show that the nonlinear operator B is still continuous; then, by virtue of the compactness of L^{-1}, the composition $L^{-1}B$ is completely continuous on the ball $\overline{B}_R$.

Assume that there is an element $h \varepsilon H$ and a sequence $\{x_n\}$ in H for which

$$\lim_{n\to\infty} ||h - x_n|| = 0, \quad ||Bh - Bx_n||^2 \geq \eta > 0 \tag{22}$$

Let $E = [0,2\pi]$ and M be a measurable subset of E; let, for abbreviation, $(Bh)(t) = k(t)$, $(Bx_n)(t) = y_n(t)$. Then we have

$$|k(t)-y_n(t)| \leq \beta|h(t)-x_n(t)| + 2\beta|h(t)| + 2P \quad \text{a.e.}$$

and

$$\int_M (k(t) - y_n(t))^2 dt < \frac{\eta}{2} \text{ for all } n \text{ if } |M| < \mu \tag{23'}$$

due to the absolute continuity of the Lebesgue integral. From (22) it follows that $\lim_{n\to\infty} x_n(t) = h(t)$ a.e. and, by continuity of $g(x)$, $\lim_{n\to\infty} y_n(t) = k(t)$ a.e. According to Egorov's theorem the subset M can be chosen in such a way that $\lim_{n\to\infty}[y_n(t)-k(t)] = 0$ uniformly on E - M. Hence,

$$\lim_{n\to\infty} \int_{E-M} (y_n(t)-k(t))^2 dt = 0 \tag{23''}$$

The results (23') and (23") are in contradiction to the second part of (22). Thus the operator B is continuous.

Choosing

$$R \geq \alpha\beta R + \alpha||y||$$

we find that the image of $\overline{B}_R$ under the completely continuous mapping

$$x \to L^{-1}Bx + L^{-1}y$$

is within $\bar{B}_R$. According to Schauder's theorem there is at least one solution of (21) with $||x|| \leq R$.

THE GENERALIZED LIENARD EQUATION

The generalized Liénard equation

$$x'' + f(x)x' + g(x) = e(t) \equiv e(t+2\pi) \tag{24}$$

where f(x), g(x), and e(t) are continuous functions has been studied extensively in some recent papers (see, for instance, [4],[10],[12],[13]). Admitting a forcing term with vanishing mean value, i.e.

$$\int_0^{2\pi} e(t)dt = 0,$$

Lazer [4] and Mawhin [10] considered the case when the restoring term is sublinear:

$$0 \leq \frac{g(x)}{x} \to 0 \quad (X \leq |x| \to \infty) \tag{25}$$

The damping coefficient f(x) was assumed to be an arbitrary constant or an arbitrary function, respectively.

By means of the Leray-Schauder topological degree the following extension of these results can be proved ([12],[13]). There is at least one 2π-periodic solution of equation (24) when

A. $xg(x) \leq 0 \ (|x| \geq X)$, $\int_0^{2\pi} e(t)dt = 0$, or

B. $0 \leq q \leq \frac{g(x)}{x} \leq p < 1 \ (|x| \geq X)$, $\int_0^{2\pi} e(t)dt = 0$ in case $q = 0$.

Let us outline the proof in case B which is more important than case A. Consider a more general differential equation

$$x'' + px = \mu\{e(t) + px - g(x) - (F(x))'\}, \quad 0 \leq \mu \leq 1 \tag{26}$$

where the bound p is assumed to be positive and where $F(x) = \int_0^x f(s)ds$. Using the Green function $\gamma(t)$ of example (13) (where $c = 0$, $\nu = p$, $m = 0$) we can represent the 2π-periodic solutions of (26) as the continuous solutions of an integral equation of Hammerstein type:

$$\begin{aligned} x(t) &= \mu T\, x(t) \\ &= \mu\int_0^{2\pi} \gamma(t-s)\{e(s) + px(s) - g(x(s))\}ds \\ &- \mu\int_0^{2\pi} \gamma'(t-s)F(x(s))ds \end{aligned} \tag{27}$$

The operator T is considered as a mapping of a suitable Banach space

$$X = \{x(t) \in C^o(R,R) : x(t+2\pi) \equiv x(t),\ ||x|| = \max_t |x(t)|\}$$

into itself. This mapping is bounded as well as completely continuous on each bounded subset of X. A fixed point under T corresponds to a 2π-periodic solution of the original equation (24).

If there is a bounded open subset $\Omega \subset X$ with the property

$$(id_X - \mu T)x \neq 0 \text{ for all } x \in \partial\Omega \ (0 \leq \mu \leq 1)$$

the Leray - Schauder degree of the mapping $id_X - \mu T$ with respect to Ω and to the zero vector is defined : $d[id_X-\mu T,\Omega,0]$. Applying the homotopy theorem according to which this mapping degree is independent of μ we obtain that $d[id_X - T, \Omega, 0] = d[id_X, \Omega, 0]$. Taking account of the fact that $d[id_X, \Omega, 0] = 1$ if and only if $0 \in \Omega$, and that $d[id_X - T, \Omega, 0] \neq 0$ implies there exists $x \in \Omega$ such that $(id_X - T)x = 0$, we can solve the problem of periodic solutions, for instance, by constructing a ball $B_R \subset X$ with a sufficiently large radius B such that $(id_X - \mu T)x \neq 0$ for all $x \in \partial B_R$. This could be done by means of the proof that all 2π- periodic solutions of equation (26) admit an a priori bound which is the same for all parameter values $\mu \in [0,1]$.

Note that a restriction to $\mu \in [0,1)$ is possible since in the case $(id_X - T)x = 0$ for an $x \in \partial B_R$ no further argumentation is needed.

The boundedness results from two simple lemmas concerning the oscillatory behavior of the solutions of (26) (see [13]).

The proofs of these lemmas include, as an essential part, some estimates of the L^2-norms of x and of x' (or of expressions composed of x and of x') corresponding to certain subintervals [a,b].

LEMMA 1. Let $x(t)$, $a \leq t \leq b$ $(b-a \leq 2\pi)$, be a solution of (26), let $x(t) \geq 0$ (≤ 0) for all $t \in [a,b]$. Moreover, let (i) $x(a) = x(b) = 0$, $|x'(a)-x'(b)| \leq \eta'$ or (ii) $x(a) = x(b)$ with $|x(a)| \leq \eta$, $x'(a) = x'(b)$. Then there is an estimate

$$|x(t)| \leq L_o(\eta, \eta') \text{ for all } t \in [a,b] \tag{28}$$

where the bound L_o is independent of μ and where $\eta = 0$ in case (i), $\eta' = 0$ in case (ii).

LEMMA 2. Let $x(t)$, $a \leq t \leq b$ $(b-a \leq \pi)$, be a solution of (26), and let $x(a) = x(b) = 0$. Then

$$|x(t)| \leq K_o, \; |x'(t)| \leq K_1 \text{ for all } t \in [a,b] \tag{29}$$

where the bounds are system constants, independent of μ.

Consider, on the basis of these statements, a 2π-periodic solution $x(t)$ of equation (26) (fixed point under μT), $0 \leq \mu < 1$.

Assume that this solution is non-oscillatory: $x(t) \neq 0$ for all $t \in R$. Integrating differential equation (26) from 0 to 2π and taking account of the periodicity we find that

$$(1 - \mu)\, p\int_0^{2\pi} x(t)dt + \mu\int_0^{2\pi} [g(x(t)) - e(t)]dt = 0$$

Consequently, $|x(t)| > X$ for all $t \in R$ is excluded: $q = 0$, $\int_0^{2\pi} e(t)dt = 0$, $(1 - \mu)p|x| + \mu g(x)\operatorname{sgn} x > 0$ $(|x| \geq X)$; $q > 0$, $(1 - \mu)p|x| + \mu[g(x)\operatorname{sgn} x - ||e||] > 0$ $(|x| \geq X > q^{-1}||e||)$. Thus, $|x(\tau)| < X$ for some $\tau \in [0,2\pi]$; applying Lemma 1 where $a = \tau$, $b = \tau + 2\pi$; $\eta = X$, $\eta' = 0$ (case (ii)) we obtain a uniform bound for the considered type of periodic solutions.

Now assume that the periodic solution $x(t)$ is oscillatory and that there is an interval [a,b], $\pi < b-a \leq 2\pi$ where $x(a) = x(b) = 0$, but $x(t) \neq 0$ on (a,b). If $b - a = 2\pi$, Lemma 1 yields that $|x(t)| \leq L_o(0,0)$ for all t. If $b - a < 2\pi$, Lemma 2 is applied to the intervals $[b-2\pi,a]$, $[b,a+2\pi]$; so $|x(t)| \leq K_0$ and $|x'(t)| \leq K_1$. After that, Lemma 1 is applied to the interval $[a,b]$; so $|x(t)| \leq L_0(0,\eta')$ with $\eta' = 2K_1$.

Finally, assume that the periodic solution $x(t)$ is oscillatory and that $x(t) \neq 0$ on (a,b), $x(a)=x(b)=0$ implies $b-a \leq \pi$. Let $x(\tau_0) = 0$ and $\tau_1 = \sup\{t : \tau_0 < t \leq \tau_0 + \pi,\ x(t) = 0\}$, $\tau_2 = \inf\{t : \tau_0 + \pi \leq t < \tau_0 + 2\pi,\ x(t) = 0\}$; then $x(\tau_1) = x(\tau_2) = 0$. Applying Lemma 2 to the intervals $[\tau_0,\tau_1]$, $[\tau_1,\tau_2]$ and $[\tau_2,\tau_0+2\pi]$ (in case $\tau_1 < \tau_2$) or to the intervals $[\tau_0,\tau_0+\pi]$ and $[\tau_0+\pi,\tau_0+2\pi]$ (in case $\tau_1 = \tau_2$) we obtain that $|x(t)| \leq K_0$ for all t.

<u>Note.</u> The generalized Liénard equation (24) admits a 2π-periodic solution, too, when $q \leq \frac{g(x)}{x} \leq p < 1$ $(pq < 0)$, and $\{r \in R : g(x)\operatorname{sgn} x < -||e|| \text{ if } |x| = r\}$ is an unbounded set.

THE VECTOR LIENARD EQUATION

According to our initial remarks (see the Introduction) we consider equation (1) under the following simplified conditions: $C = \operatorname{diag}(c_1,\ldots,c_n)$, $(N - H(x))^2 \leq (P - Q)^2/4$, $N = (P + Q)/2$, $P = \operatorname{diag}(p_1,\ldots,p_n)$, $Q = \operatorname{diag}(q_1,\ldots,q_n)$ and (i) $m_r^2 < q_r \leq p_r < (m_r+1)^2$, m_r non-negative integer, or, (ii) $q_r \leq p_r < 0$. Introducing the matrices $M = \operatorname{diag}(\mu_1,\ldots,\mu_n)$ where (i) $\mu_r = m_r^2$ or (ii) $\mu_r < 0$, and $M' = \operatorname{diag}(\mu_1',\ldots,\mu_n')$ where (i) $\mu_r' = (m_r+1)^2$ or (ii) $\mu_r' = 0$, we can assume, without loss of generality, that $p_r = \mu_r' - \varepsilon(\mu_r'-\mu_r)/2$, $q_r = \mu_r + \varepsilon(\mu_r'-\mu_r)/2$ $(0 < \varepsilon < 1,\ \varepsilon$ sufficiently small). Then $N = \operatorname{diag}(\nu_1,\ldots,\nu_n)$, $\nu_r = (\mu_r'+\mu_r)/2$ and

$$(N - H(x))^2 \leq (1 - \varepsilon)^2\,(M' - M)^2/4 \tag{30}$$

Hence $||N - H(x)|| = \sup\{||(N - H(x))e|| : e \in R^n,\ ||e|| = 1\} \leq \frac{1-\varepsilon}{2}||M' - M||$. An immediate consequence of this estimate is the linear boundedness of the vector field $Nx - \operatorname{grad} G(x)$:

$$||Nx - \operatorname{grad} G(x)|| \leq k_1||x|| + k_0 \tag{31}$$

In order to generalize the periodic boundary value problem of the vector differential equation (1) in a similar way to the above, we make use of the previous results concerning the operator $L = L_r$ of equation (13) where $c = c_r$ and $\nu = \nu_r$ namely that there is a bounded inverse, and

$$||L_r^{-1}|| \le \alpha_r = \frac{2}{\mu_r' - \mu_r} \tag{32}$$

Consider the (real) Hilbert space $\hat{H}$ of all $x(t) = \mathrm{col}(x_1(t), \ldots, x_n(t))$ with $x_r(t) \varepsilon H$ $(1 \le r \le n)$; let us introduce the inner product

$$<x,y>_{\hat{H}} = \sum_{r=1}^{n} \alpha_r^{-2} <x_r, y_r>_H$$

Moreover, consider the subspace $\hat{D}$ which consists of all $x \varepsilon \hat{H}$ the components of which belong to $D = D(L)$. Define on $\hat{D}$,

$$\hat{L}x = \mathrm{col}(L_1 x_1, \ldots, L_n x_n) \tag{33}$$

The operator $\hat{L}$ maps $\hat{D}$ onto $\hat{H}$, and it has a bounded inverse, $\hat{L}^{-1}y = \mathrm{col}(L_1^{-1}y_1, \ldots, L_n^{-1}y_n)$. Taking account of (31) we define a nonlinear operator on $\hat{H}$ by

$$\begin{aligned}\hat{B}x &= Nx - \mathrm{grad}\, G(x) \\ &= \mathrm{col}(\nu_1 x_1(t) - G_{x_1}(x(t)), \ldots, \nu_n x_n(t) - G_{x_n}(x(t))) \\ &\quad (\mathrm{a.e.})\end{aligned} \tag{34}$$

At each point $x \varepsilon \hat{H}$ it has a Gateaux derivative $\hat{B}'(x)$ which is a linear operator on $\hat{H}$, defined by

$$\begin{aligned}(\hat{B}'(x)u)(t) &= (N - H(x))u(t) \\ &= \mathrm{col}\,(\nu_r u_r(t) - \sum_{s=1}^{n} h_{rs}(x(t))u_s(t))\ (\mathrm{a.e.})\end{aligned}$$

where $u \varepsilon \hat{H}$ arbitrary, $H(x) = (h_{rs}(x))$. Note that $\hat{L}^{-1}\hat{B}$ has the Gâteaux derivative $\hat{L}^{-1}\hat{B}'(x)$. Furthermore, note that the mean value theorem (see [14]) can be applied in order to derive the estimate

$$||\hat{L}^{-1}\hat{B}u - \hat{L}^{-1}\hat{B}v|| \le ||\hat{L}^{-1}\hat{B}'(\hat{x})||\ ||u - v|| \tag{35}$$

($\hat{x} \varepsilon \hat{H}$ on the segment of extremities u, v).

Now we replace the differential equation (1) with periodic boundary conditions by the operator equation

$$\hat{L}x - \hat{B}x = y \quad (x \varepsilon \hat{D},\ y \varepsilon \hat{H})$$

or, equivalently,

$$x = \hat{L}^{-1}\hat{B}x + \hat{L}^{-1}y \tag{36}$$

It can easily be shown that the operator $\hat{L}^{-1}\hat{B}$ is a contraction in $\hat{H}$. For this purpose we apply (35) and we derive a global estimate for $||\hat{L}^{-1}\hat{B}'(\hat{x})||$:

$$||\hat{L}^{-1}\hat{B}'(\hat{x})w||_{\hat{H}}^2 = \sum_{r=1}^{n} \alpha_r^{-2}||L_r^{-1}\{\sum_{s=1}^{n} \eta_{rs}(t)w_s(t)\}||_H^2$$

$$\leq (2\pi)^{-1}\int_0^{2\pi} \sum_{r,s=1}^{n} \eta_{rs}^{(2)}(t)w_r(t)w_s(t)dt$$

where, for abbreviation, $N\text{-}H(\hat{x}) = (\eta_{rs}(t))$, $(N\text{-}H(\hat{x}))^2 = (\eta_{rs}^{(2)}(t))$. Taking account of (30) we obtain

$$\sum_{r,s=1}^{n} \eta_{rs}^{(2)}(t)w_r(t)w_s(t) \leq \frac{1}{4}(1-\varepsilon)^2 \sum_{r=1}^{n} (\mu_r'-\dot{\mu}_r)^2 w_r^2(t)$$

$$= (1-\varepsilon)^2 \sum_{r=1}^{n} \alpha_r^{-2} w_r^2(t)\,(\text{a.e.})$$

from which we conclude that $||\hat{L}^{-1}\hat{B}'(\hat{x})w||_{\hat{H}}^2 \leq (1-\varepsilon)^2||w||_{\hat{H}}^2$ and $||\hat{L}^{-1}\hat{B}'(\hat{x})|| \leq 1-\varepsilon$. Hence, Banach's fixed point theorem is applicable in order to solve equation (36). Again, the successive approximations and their first derivatives converge uniformly on $[0,2\pi]$. Using a 2π-periodic continuous function $e(t)$ we obtain the classical periodic solution of equation (1).

REFERENCES

1. S. Ahmad, An existence theorem for periodically perturbed conservative systems, Michigan Math. J. 20 (1973), 385-392.
2. E. Hewitt and K. Stromberg, Real and Abstract Analysis, Springer-Verlag, Berlin, 1969.
3. R. Kannan, Periodically perturbed conservative systems, J. Differential Equations 16 (1974), 506-514.
4. A. C. Lazer, On Schauder's fixed point theorem and forced second-order nonlinear oscillations, J. Math. Anal. Appl. 21 (1968), 421-426.
5. A. C. Lazer, Application of a lemma on bilinear forms to a problem in nonlinear oscillations, Proc. Amer. Math. Soc. 33 (1972), 89-94.

6. A. C. Lazer and D. A. Sanchez, On periodically perturbed conservative systems, Michigan Math. J. 16 (1969), 193-200.

7. D. E. Leach, On Poincaré's perturbation theorem and a theorem of W. S. Loud, J. Differential Equations 7 (1970), 34-53.

8. W. S. Loud, Periodic solutions of nonlinear differential equations of Duffing type, Proc. U. S.-Japan Sem. on Diff. and Funct. Equations, Benjamin, New York, 1967, 199-224.

9. K. Löwner, Uber monotone Matrixfunktionen, Math. Z. 38 (1934), 177-216.

10. J. Mawhin, An extension of a theorem of A. C. Lazer on forced nonlinear oscillations, J. Math. Anal. Appl. 40 (1972), 20-29.

11. J. Mawhin, Contractive mappings and periodically perturbed conservative systems, Sém. Math. Appl. Méc., rapport 80, Louvain, 1974.

12. R. Reissig, Extension of some results concerning the generalized Liénard equation, Ann. Mat. Pura Appl. 104 (1975), 269-281.

13. R. Reissig, Schwingungssätze für die verallgemeinerte Liénardsche Differentialgleichung, Abh. Math. Sem. Uni. Hamburg 44 (1975), 45-51.

14. M. M. Vainberg, Variational Methods for the Study of Nonlinear Operators, Holden-Day, San Francisco, 1964.

Chapter 5

AN OSCILLATION THEOREM FOR A HIGHER ORDER DIFFERENTIAL EQUATION

R. REISSIG

Institut für Mathematik
Ruhr-Universität Bochum
Federal Republic of Germany

This chapter is devoted to the n-th order vector differential equation

$$L[x^{(k)}] + f(x) \equiv x^{(n)} + A_1 x^{(n-1)} + A_2 x^{(n-2)} + \dots + A_{n-k-1} x^{(k+1)} + A_{n-k} x^{(k)} + f(x) = e(t) \qquad (1)$$

where $n \geq 2$, $1 \leq k \leq n-1$, $x \in R^m$ $(m \geq 1)$, $f : R^m \to R^m$ continuous, $e : R \to R^m$ continuous and 2π-periodic, $A_\nu (0 \leq \nu \leq n-k)$ a real constant $m \times m$-matrix, $A_o = I_m$. Note that the characteristic polynomial belonging to the linear differential operator L is λ^{mk} $p(\lambda)$ where

$$p(\lambda) = \det P(\lambda),\ P(\lambda) = \sum_{\nu=k}^{n} \lambda^{\nu-k} A_{n-\nu}$$

Let us introduce some abbreviations:

$$\max|e(t)| = E^*,\ \int_0^t e(s)ds = E(t)[2\pi\text{-periodic if } E(2\pi) = 0]$$

$$\phi(r) = \sup\{|f(x)| : |x| \leq r\}$$

$$\psi(r) = \inf\{\langle x, f(x)\rangle : |x| = r\}$$
$$f_\mu(x) = (1+\mu)f(x)/2 - (1-\mu)f(-x)/2, \; 0 \le \mu \le 1$$
$$\phi_\mu(r) = \sup\{|f_\mu(x)| : |x| \le r\} \le \phi(r)$$
$$\psi_\mu(r) = \inf\{\langle x, f_\mu(x)\rangle : |x| = r\} \ge \psi(r)$$

The aim of our chapter is to develop conditions which ensure the existence of at least one 2π-periodic solution. In the special case $k = 1$ there are some recent results of Sedziwy, Mawhin and of other authors (see [1-5]) which shall be extended partially. Applying Brouwer's fixed point theorem on the basis of boundedness results, Sedziwy stated the following theorem concerning the oscillatory properties of equation (1).

THEOREM 1. ([4]) Assume that (i) $p(\lambda) = 0$ implies that $\mathrm{Re}\lambda < 0$, (ii) $A_{n-1} = A_{n-1}^* > 0$, (iii) $|f(x)| \le F^*$, and (iv') $\lim_{|x|\to\infty} \langle x, f(x)-q\rangle = \infty$ for all $q \in R^m$ such that $|q| \le E^*$, or (iv'') $\lim_{|x|\to\infty} \langle x, f(x)\rangle = \infty$, $E(2\pi) = 0$. Then there exists at least one 2π-periodic solution of (1).

Another result of Sedziwy was derived by means of Borsuk's theory on odd mappings of finite dimensional spaces.

THEOREM 2. ([5]) Assume that (i) $p(\lambda) = 0$ implies that $\lambda \ne ri$ (r an arbitrary integer), (ii) $A_{n-1} = A_{n-1}^* > 0$, (iii) $|f(x)| \le \varepsilon|x|$ for all $x \in R^m - B_h$, (iv) $\langle x, f(x)\rangle \ge \delta|x||f(x)|$, $0 < \delta < 1$ for all $x \in R^m$, and (v') $\inf\{|f(x)| : x \in R^m - B_h\} > \delta^{-1}E^*$, or (v'') $E(2\pi) = 0$. Then there is at least one periodic solution provided that $\varepsilon > 0$ is small enough.

Using the functional analytic method based on completely continuous operators in certain Banach spaces Mawhin proved a rather general result; however, his condition on the nonlinear restoring term f concerns its components whereas a condition for the inner product $\langle x, f(x)\rangle$ seems to be more natural.

THEOREM 3. ([1,2]) Assume that (i) $p(\lambda) = 0$ implies that $\lambda \ne ri$ ($r \ne 0$ an arbitrary integer), (ii) $\lim_{|x|\to\infty} [|f(x)|/|x|] = 0$, (iii) $\sigma_\mu x_\mu f_\mu(x) \ge 0$ if $|x_\mu| \ge h_\mu$ $(1 \le \mu \le m)$ where $\sigma_\mu = +1$ or

-1, and (iv) $E(2\pi) = 0$. Then equation (1) admits at least one periodic solution.

Now let us formulate our statement.

THEOREM 4. The following conditions are sufficient for the existence of a periodic solution: (i) $p(\lambda) = 0$ implies that $\lambda \neq ri$ (r an arbitrary integer); (ii) $A_{n-1} = A^*_{n-1}$ (in case $k = 1$); (iii) $E(2\pi) = 0$; there is some domain $\{x \in R^m : H \leq |x| \leq 2H$, H large enough$\}$ where (iv') $\phi(r) \leq \beta r^\alpha$ $(0 \leq \alpha < 1)$, $\psi(r) \geq Kr^{2\alpha}$ (K large enough), or (iv") $\phi(r) \leq 1 + \varepsilon r^\alpha$ $(0 < \alpha \leq 1$, ε small enough if $\alpha = 1)$, $\psi(r) \geq \varepsilon r^{2\alpha}$.

COROLLARY. If n is odd and $A_\nu = 0$ for ν odd, $A_\nu = A_\nu^*$ for ν even, or, if n a multiple of 4 and $A_\nu = 0$ for ν even, $A_\nu = A_\nu^*$ for ν odd, then $r^{2\alpha}$ can be replaced by r^α in the estimate for $\psi(r)$.

The proof of Theorem 4 is based on the Leray-Schauder fixed point theory and on the fundamental theorem on odd mappings of Banach spaces. To begin with we choose a sufficiently small positive number a such that the polynomial $p_a(\lambda) = \det(\lambda^k p(\lambda) + aI_m)$ has no zero $\lambda = ri$ (r an integer). This is possible since $p_o(\lambda) = \lambda^{mk}p(\lambda) = 0$ implies that either $\lambda = 0$ or $p(\lambda) = 0$ (which means that $\lambda \neq ri$), $p_a(0) = a^m > 0$, and since the roots of $p_a(\lambda)$ are continuously depending upon the parameter value a. Now we consider the auxiliary equation

$$L[x^{(k)}] + ax = \mu e(t) + ax - f_\mu(x), \quad 0 \leq \mu \leq 1 \tag{2}$$

where $f_\mu(x)$ is defined above. Note that $f_o(-x) = -f_o(x)$, $f_1(x) = f(x)$ and that $f_\mu(x)$ satisfies conditions (iv') or (iv"). The periodic solutions of (2) can be represented as

$$x(t) = \theta_\mu[x(t)] = \int_0^{2\pi} G(t-s)[\mu e(s)+ax(s)-f_\mu(x(s))]ds \tag{3}$$

where G is the Green's matrix belonging to the linear differential operator $L[x^{(k)}] + ax$ in connection with periodic boundary conditions $x^{(\nu)}(2\pi) = x^{(\nu)}(0)$ $(0 \leq \nu \leq n-1)$.

Introducing the Banach space

$$X = \{x(t) \in C^o(R) : x(t+2\pi) \equiv x(t)\}, \quad ||x||_X = \max|x(t)|$$

we can state that the mapping

$$x \to \theta_\mu(x),\ \mu \in [0,1] \text{ fixed}$$

is completely continuous on each bounded subset of X. The parameter μ describes a continuous deformation. If there is a ball $B_R = \{x \in X : ||x|| < R\}$ such that $(id_X - \theta_\mu)\, x \neq 0$ on ∂B_R for all $\mu \in [0,1]$ then the Leray-Schauder degree $d[id_X - \theta_\mu, B_R, 0]$ is independent of μ. It is an odd integer since the mapping $\theta_o(x)$ is odd. Consequently, the equation $(id_X - \theta_\mu)\, x = 0$ has at least one solution $x \in B_R$. In case $\mu = 1$ this solution satisfies equation (1) under periodic boundary conditions. Let us show that B_R for R = 2H (according to conditions (iv') or (iv")) is such a ball. For this purpose assume that

$$x(t) = \theta_\mu[x(t)],\ 0 \leq \mu \leq 1;\ ||x||_X = 2H,\ \min|x(t)| = h$$

We are looking for an estimate of the derivatives $y^{(\nu)} = x^{(k+\nu)}$, $0 \leq \nu \leq n-1-k$. Since

$$L[y] = \mu e(t) - f_\mu(x(t))$$

$$y^{(\nu)}(2\pi) = y^{(\nu)}(0)\ \ (0 \leq \nu \leq n-1-k) \tag{4}$$

where the "homogeneous" problem Ly = 0 only admits the zero solution, the derivatives $y^{(\nu)}(t)$ can be represented in integral form by means of a Green's matrix which is, at least, piecewise continuous. As an immediate consequence we can estimate

$$||y^{(\nu)}||_X \leq \rho_{k+\nu}[E^* + \phi(2H)],\ 0 \leq \nu \leq n-k-1 \tag{5}$$

In case k > 1 we consider $x^{(k-1)}$. On an arbitrary interval with extremities t_o and $t_o + 2\pi$ we have

$$|x^{(k-1)}(t) - x^{(k-1)}(t_o)| \leq 2\pi\rho_k[E^* + \phi(2H)]$$

This estimate is valid, too, for every single component; using a zero t_o of the component and taking into account the m components we find that

$$||x^{(k-1)}||_X \leq \rho_{k-1}[E^* + \phi(2H)]$$

Continuing this procedure, if necessary, we obtain the result

$$||x^{(\nu)}||_X \leq \rho_\nu[E^* + \phi(2H)],\ 1 \leq \nu \leq k-1 \tag{6}$$

Integration of the inner product between x(t) and both members of equation (2) yields $\int_0^{2\pi} <x, x^{(n)} + A_1 x^{(n-1)} + \ldots + A_{n-k} x^{(k)} - \mu E'>dt + \int_0^{2\pi} <x, f_\mu(x)>dt = \int_0^{2\pi} <x, f_\mu(x)>dt - \int_0^{2\pi} <x', x^{(n-1)} + A_1 x^{(n-2)} + \ldots + A_{n-k} x^{(k-1)} - \mu E>dt = 0.$

Using the fact that in case k = 1

$$\int_0^{2\pi} <x', A_{n-1}x>dt = (1/2)\int_0^{2\pi} <x, A_{n-1}x>'dt = 0$$

we conclude, with the aid of (5) and (6),

$$\int_0^{2\pi} <x, f_\mu(x)>dt \le \rho[E^* + \phi(2H)]^2 \tag{7}$$

The constant ρ only depends on the characteristics of the differential operator L.

<u>Note.</u> In the special cases described in the Corollary the estimate (7) can be improved as follows:

$$\int_0^{2\pi} <x, f_\mu(x)>dt \le 2\pi\rho_1 E^*[E^* + \phi(2H)] \tag{8}$$

Let $|x(t_1)| = h$ and $|x(t_2)| = 2H$ where $|t_1 - t_2| \le 2\pi$. Then $2H - h \le |x(t_1) - x(t_2)| \le 2\pi\rho_1[E^* + \phi(2H)]$, $h/H \ge 2 - 4\pi\rho_1[E^* + \phi(2H)]/2H \ge 1$, i.e. $H \le h$ if $H \ge 4\pi\rho_1 E^*$, $\phi(2H)/2H \le (8\pi\rho_1)^{-1}$ which is ensured, under conditions (iv') or (iv"), whenever H is large enough and (in case $\alpha = 1$) $\varepsilon > 0$ is small enough.

Taking into account the last result and assuming condition (iv') holds, we derive from (7)

$$K \le (1/2\pi) \int_0^{2\pi} [<x, f_\mu(x)>/|x|^{2\alpha}]dt$$

$$\le 2\rho\{[E^* + \phi(2H)]/(2H)^\alpha\}^2/\pi$$

$$\le 2\rho[E^*/(2H)^\alpha + \beta]^2/\pi$$

$$\le 2\rho(E^* + \beta)^2/\pi$$

provided that $H \geq 1$. This is a contradiction when K is large enough. Assuming condition (iv") holds, we calculate

$$\varepsilon \leq 2\rho[(E^*+1)/(2H)^{\alpha} + \varepsilon]^2/\pi \leq 8\rho\varepsilon^2/\pi$$

provided that $H \leq (\frac{E^*+1}{\varepsilon})^{1/\alpha}/2$. This is a contradiction when ε is small enough.

A similar argumentation can be given in the particular situation of the Corollary.

As a result we state that there is no $x(t) = \theta_{\mu}[x(t)]$, $0 \leq \mu \leq 1$, with $||x||_X = 2H$.

REFERENCES

1. J. Mawhin, Periodic solutions of some vector retarded functional differential equations, J. Math. Analys. Appl. 45 (1974), 588-603.
2. J. Mawhin, Degré de coincidence et problèmes aux limites pour des équations différentielles ordinaires et fonctionnelles, Rapport no. 64, Sém. math. appl. méc., Université Catholique de Louvain, 1973.
3. R. Reissig, Periodic solutions of a nonlinear n-th order vector differential equation, Ann. Mat. Pura Appl. 87 (1970), 111-124.
4. S. Sedziwy, Asymptotic properties of solutions of a certain n-th order vector differential equation, Atti Accad. Naz. Lincei Rend. Cl. Sci. Fis. Mat. Natur. 47 (1969), 472-475.
5. S. Sedziwy, Periodic solutions of a system of nonlinear differential equations, Proc. Amer. Math. Soc. 48 (1975), 328-336.

Part II

CONTRIBUTED PAPERS

Chapter 6

SEMI-COMPACTNESS IN SEMI-DYNAMICAL SYSTEMS

PREM N. BAJAJ
Department of Mathematics
Wichita State University
Wichita, Kansas

INTRODUCTION

Semi-dynamical systems (s.d.s.) are continuous flows defined for all future time (non-negative t). Natural examples of s.d.s.are provided by functional differential equations for which existence and uniqueness conditions hold. S.d.s. generalize the theory of Dynamical Systems. Moreover many new and interesting notions (e.g., start point [2], [3], singular point [1], [4]) arise in s.d.s.

In this chapter, an attempt is made to weaken the hypothesis on the phase space. It is the semi-compactness, rather the local compactness, that counts. After introducing basic notions, we proceed to examine some properties of limit sets, prolongations, and their limit sets. In particular, we prove that in a rim-compact space, the positive limit set and the positive prolongational limit set are weakly negatively invariant and do not contain any start points. The goal of the last section (on stability and asymptotic stability) is to prove that, in a rim-compact space, if a positively invariant closed set with compact boundary is a uniform attractor, it is asymptotically stable.

BASIC NOTIONS

Definitions

A semi-dynamical system (s.d.s.) is a pair (X,π) where X is a topological space and π is a continuous map from $X \times R^+$ into X satisfying the conditions:

$\pi(x,0) = x$, $x \in X$, (identity axiom)

$\pi(\pi(x,t),s) = \pi(x,t+s)$, $x \in X$; $t, s \in R^+$ (semi-group axiom)

(R^+ is the set of nonnegative reals with usual topology). For brevity $\pi(x,t)$ will be denoted by xt, the set $\{xt : x \in M \subset X, t \in K \subset R^+\}$ by MK etc.

For any t in R^+, the map $\pi^t : X \to X$ is defined by $\pi^t(x) = xt$. The negative funnel, $F(x)$, from x is the set $\{y \in X : yt = x$ for some $t \in R^+\}$. For any t in R^+, the function π^{-t} defined on X with values in the set of subsets of X is given by $\pi^{-t}(x) = \{y \in X : yt = x\}$. Clearly $F(x) = \cup\{\pi^{-t}(x) : t \in R^+\}$. Positive trajectory, $\gamma^+(x)$, and positive invariance are defined as in dynamical systems, [5], [6]. A negative trajectory, $\gamma^-(x)$, from any point x is a maximal non-empty subset of $F(x)$ such that for any y,z in $\gamma^-(x)$, if $t(y) = \text{Inf}\ \{s \geq 0 : ys = x\}$ and $t(z) = \text{Inf}\ \{s \geq 0 : zs = x\}$, then $y \in z[0, t(z)]$ or $z \in y[0, t(y)]$. In general there will be more than one negative trajectory from x. A negative trajectory $\gamma^-(x)$ will be called a principal negative trajectory if the set $\{t \in R^+ : yt = x$ for some y in $\gamma^-(x)\}$ is unbounded. A point $n \in X$ is said to be positive critical (or simply critical) if $\gamma^+(x) = \{x\}$.

A subset K of X is said to be negatively invariant (weakly negative invariant) if the negative funnel (at least one negative trajectory) from each point of K lies in K.

For any x in X, let $E(x) = \{t \geq 0 : yt = x$ for some y in $X\}$. The escape time of x is said to be infinite or $\text{Sup}\ E(x)$ according to whether $E(x)$ is unbounded or bounded. A point with zero escape time is said to be a start point. See [2], [3] for some results on start points.

Notation

Throughout this chapter (X,π) denotes a semi-dynamical system where X is taken to be Hausdorff. $N(x)$ denotes the neighborhood (nbd.) filter [10; p. 78] of x. A net in X will be referred to as x_i where i is in the directed set and x_i is its image.

LIMIT SETS, PROLONGATIONS AND PROLONGATIONAL LIMIT SETS

Definition

The positive limit set $\Lambda(x)$, positive prolongation $D(x)$, and positive prolongational limit set $J(x)$ of a point x in X are defined by:

$$\Lambda(x) = \{y \in X : \text{there exists a net } t_i \text{ in } R^+,\ t_i \to \infty, \text{ such that } xt_i \to y\}$$

$$D(x) = \{y \in X : \text{there exists a net } x_i \text{ in } X,\ x_i \to x, \text{ and a net } t_i \text{ in } R^+ \text{ such that } x_i t_i \to y\}$$

$$J(x) = \{y \in X : \text{there exists a net } x_i \text{ in } X,\ x_i \to x, \text{ and a net } t_i \text{ in } R^+,\ t_i \to \infty \text{ such that } x_i t_i \to y\}$$

PROPOSITION 1. Let $x \in X$. Then $\Lambda(x) = \cap\{C\ell(\gamma^+(xt)) : t \in R^+\}$.

DEFINITION. A topological space is said to be rim-compact (or semi-compact) if it has a base of open sets with compact boundaries ([9; p. 111]).

Rim-compact Hausdorff space is easily seen to be regular.

THEOREM 2. Let X be rim-compact, $x \in X$, and $\Lambda(x)$ be non-empty and compact. Then (a) $C\ell(xR^+)$ is compact, and (b) $\Lambda(x)$ is connected.

The above theorem, in effect, states that if X is rim-compact, then the compactness of $\Lambda(x)$ implies positive Lagrange Stability [5].

The following theorem, well-known for Dynamical Systems, holds for semi-dynamical systems also.

<u>THEOREM 3.</u> Let $x \in X$. Then $D(x) = \cap\{C\ell(VR^+) : V$ is a nbd. of $x\}$.

<u>THEOREM 4.</u> Let X be rim-compact and $x \in X$. Then both $\Lambda(x)$ and $J(x)$ are weakly negatively invariant and contain no start points.

<u>Proof.</u> We outline the proof for $J(x)$ only; the proof for $\Lambda(x)$ is similar.

Let $y \in J(x)$. Then there exists a net x_i in X, $x_i \to x$, and a net t_i in R^+, $t_i \to +\infty$, such that $x_i t_i \to y$. Let $t > 0$ be arbitrary but fixed. We may suppose $t_i > t$ for every i. Now consider the net $x_i(t_i - t)$ in X. If $x_i(t_i - t)$ converges to y, then $x_i(t_i - t)t \to yt$, moreover $x_i(t_i - t)t = x_i t_i \to y$, so that $yt = y$.

If $x_i(t_i - t)$ does not converge to y, there exists a nbd. V of y such that a subnet of $x_i(t_i - t)$ is in $X - V$; we may take V open, with its boundary, Fr(V), compact. For simplicity of notation, let the subnet be $x_i(t_i - t)$ itself. Since $x_i t_i \to y$, there exists an I in the directed set such that $x_i t_i \in V$ whenever $i \geq I$. Now for each $i \geq I$, there exists an s_i, $t_i - t \leq s_i < t_i$, such that $x_i s_i \in Fr(V)$. By compactness of Fr(V), $x_i s_i$ has a convergent subnet. Let $x_i s_i \to z \in Fr(V)$. Since $0 < t_i - s_i \leq t$, the net $t_i - s_i$ has a convergent subnet; let $t_i - s_i \to s$, where $0 \leq s \leq t$. Then $x_i s_i(t_i - s_i) = x_i t_i \to y$ and $x_i s_i(t_i - s_i) \to zs$, so that $zs = y$. But $z \in Fr(V)$, $y \in V$ and V is open; therefore $s \neq 0$.

In either case y is not a start point. Using the point y or z above, the existence of negative trajectory from y now follows from Hausdorff's maximality principle.

The following theorems, interesting in their own right, will be needed later.

<u>THEOREM 5.</u> Let $x \in X$, $\Lambda(x) \neq \phi$, and $\omega \in \Lambda(x)$. Then $J(x) \subset J(\omega)$.

<u>Proof.</u> Since $\omega \in \Lambda(x)$, there exists a net t_i in R^+, $t_i \to +\infty$, such that $xt_i \to \omega$. Let $y \in J(x)$, so that there exists a net x_i in X, $x_i \to x$, and a net s_i in R^+, $s_i \to +\infty$ such that $x_i s_i \to y$. By making adjustments, we have $s_i - t_i \to +\infty$. Let

U be an open nbd. of ω. Then for some I in the directed set, $xt_i \in U$ for every $i \geq I$.

Let $xt_i \in U$ for an arbitrary but fixed $i \geq I$. Since $x_j t_i \to xt_i$ whenever $x_j \to x$ and i is held fixed, we can pick $j = j(i) \geq i$ such that $x_{j(i)}t_i \in U$. Now the net $x_{j(i)}t_i$ converges to ω. Moreover $x_{j(i)}t_i(s_{j(i)} - t_i) = x_{j(i)}s_{j(i)} \to y$ and $(s_{j(i)} - t_i) \to \infty$. Hence $y \in J(\omega)$, and so $J(x) \subset J(\omega)$.

THEOREM 6. Let X be rim-compact, $x \in X$, and let $J(x)$ be non-empty and compact. Then $\Lambda(x)$ is non-empty (and compact).

NOTATION. For a subset M of X, let $D(M) = \cup\{D(x) : x \in M\}$.

For a closed set M with compact boundary, we have the following theorems.

THEOREM 7. Let M be a closed subset of X with compact boundary, and let $y \in D(x) - M$ for some x in M. Then $y \in D(z)$ and $z \in D(x)$ for some $z \in Fr(M)$.

Proof. If $x \in Fr(M)$, let $z = x$. If $x \in Int(M)$, there exists a net x_i in $Int(M)$, $x_i \to x$, and a net t_i in R^+ such that $x_i t_i \notin M$ and $x_i t_i \to y$. For each i there exists an s_i, $0 < s_i < t_i$ such that $x_i s_i \in Fr(M)$. By compactness of $Fr(M)$, $x_i s_i$ has a convergent subnet. With no loss in generality we can let $x_i s_i \to z \in Fr(M)$. Clearly $y \in D(z)$ and $z \in D(x)$.

THEOREM 8. Let M be a closed subset of X with compact boundary. Then (a) $D(M) = \cap\{C\ell(UR^+) : U \text{ is a nbd. of } M\}$, and (b) $D(M) = \cap\{U : U \text{ is a closed, positively invariant nbd. of } M\}$. In particular, it follows that $D(M)$ is closed.

Proof. To prove (a) let $K = \cap\{C\ell(UR^+) : U \text{ is a nbd. of } M\}$. For any $x \in M$, $D(x) = \cap\{C\ell(VR^+) : V \text{ is a nbd. of } M\}$. Since a nbd. of M is a nbd. of x, it follows that $D(x) \subset K$. As $x \in M$ is arbitrary, $D(M) \subset K$.

Next let $y \notin D(M)$. In particular $y \notin D(x)$ for every $x \in Fr(M)$. Then for each $x \in Fr(M)$, there exists a nbd. U_x of x, a nbd. V_x of y such that $V_x \cap U_x R^+ = \phi$. Let $\{U_1, U_2, \ldots, U_n\}$ be a finite subcover of the open cover $\{U_x : x \in Fr(M)\}$ of the compact set $Fr(M)$. Let $V = V_1 \cap \ldots \cap V_n$ and $W = U_1 \cup \ldots \cup U_n$.

Clearly $V \cap (M \cup W)R^+ = \phi$ and, so, $y \notin C\ell\ (M \cup W)R^+$. But $M \cup W$ is a nbd. of M. Hence $y \notin K$ etc.

The proof of (b) part is similar.

STABILITY, ATTRACTION AND ASYMPTOTIC STABILITY

DEFINITION. Let M be a closed subset of X with compact boundary. Then M is said to be stable if given a nbd. U of M, there exists a nbd. V of M such that $VR^+ \subset U$.

THEOREM 9. Let M be a closed subset of X with compact boundary. If M is stable, then $D(M) = M$. If X is rim-compact, the converse also holds.

Proof. We outline only the proof of the converse. Let U be a nbd. of M. We may take U to be open and with a compact boundary. It is easy to see that for each $x \in Fr(M)$, there exists a nbd. W_x of x such that $W_xR^+ \subset U$. Let $\{W_1, \ldots, W_n\}$ be a subcover of the open cover $\{W_x : x \in Fr(M)\}$ of Fr(M). Let $V = W_1 \cup \ldots \cup W_n \cup M$. Clearly $VR^+ \subset U$.

DEFINITION. Let M be a closed subset of X with compact boundary. The region of attraction, A(M), of M is defined to be the set $\{x \in X : \phi \neq \Lambda(x) \subset M\}$. M is said to be an attractor if A(M) is a nbd. of M. If M is an attractor, it is said to be a uniform attractor if given a compact set $K \subset A(M)$, and a nbd. V of M, there exists $T \geq 0$ such that $x[T, +\infty] \subset V$ for each x in K. Finally M is said to be asymptotically stable if it is stable and is an attractor.

THEOREM 10. Let M be a closed subset of X with compact boundary. Let M be positively invariant. (a) If M is asymptotically stable, $J(A(M)) \subset M$. (b) If X is rim-compact, and there exists a nbd. U of M such that $\phi \neq J(x) \subset M$ for each x in U, then M is asymptotically stable.

THEOREM 11. Let X be rim-compact, M be a closed subset of X with compact boundary, and let M be positively invariant. If M is a uniform attractor, then it is asymptotically stable.

Proof. We need show that M is stable. Let V be a neighborhood of M. We may take V to be open, Fr(V) to be compact, and $C\ell\ V \subset A(M)$.

For the compact set Fr(V), let $T > 0$ be of the definition of uniform attraction. For $x \in Fr(V)$, let $\tau(x) = \text{Inf}\ \{t \geq 0 : x[t,\infty) \subset V\}$. Since M is a uniform attractor, $\tau(x)$ is well defined and $\tau(x) \leq T$. Clearly $x\tau(x) \in Fr(V)$. Moreover, $x[0,\tau(x)] \cap M = \phi$. (This follows from the observation that M is positively invariant). Consider $F = \cup\{x[0,\tau(x)] : x \in Fr(V)\}$. We assert that F is compact. Let $x_i t_i$, $x_i \in Fr(V)$, be any net in F, $0 \leq t_i \leq \tau(x_i) \leq T$; t_i being bounded, has a convergent subnet. With no loss in generality, we let $t_i \to t$. Similarly, let $\tau(x_i) \to \tau$ so that

$$0 \leq t \leq \tau \leq T \qquad (*)$$

The net x_i, being in a compact set Fr(V), has a convergent subnet. As before, let $x_i \to x \in Fr(V)$. Now $x_i t_i \to xt$. We have to show that $xt \in F$, i.e., $t \leq \tau(x)$. Now $x_i\tau(x_i) \to x\tau \in Fr(V)$ (notice that Fr(V) is closed). Therefore $\tau \leq \tau(x)$. Hence, by (*), $0 \leq t (\leq \tau) \leq \tau(x)$, and so $xt \in X[0, \tau(x)] \subset F$. This proves the compactness of F. Since X is T_2, F is closed. Next $F \cap M = \phi$ as $F \cap x[0,\tau(x)] = \phi$ for each $x \in Fr(V)$. Let $U = V - F$. Then U is a nbd. of M and $UR^+ \subset V$. Hence M is stable.

Remarks. The proof is significant for many reasons. We have avoided the local compactness condition on the space X, and $\tau(x)$ is not necessarily continuous, as can easily be seen. Moreover, the proof is constructive in the sense that not only stability of M is established, but also the neighborhood U corresponding to the given neighborhood V is actually found such that $UR^+ \subset V$. Moreover, the neighborhood U we found is the LARGEST such neighborhood.

REFERENCES

1. P. N. Bajaj, Singular Points in Products of Semi-dynamical Systems, SIAM J. Appl. Math. 18 (1970), 282-286.

2. P. N. Bajaj, Start Points in Semi-dynamical Systems, Funkcial. Ekvac. 13 (1971), 171-177.

3. P. N. Bajaj, Connectedness Properties of Start Points in Semi-dynamical Systems, Funkcial. Ekvac. 14 (1971), 171-175.

4. P. N. Bajaj, Connectedness vs. Sets of Singular Points in Products of Semi-dynamical Systems, Math. Student 40 (1972), 329-333.

5. N. P. Bhatia and G. P. Szegö, Dynamical Systems: Stability Theory and Applications, Lect. Notes in Math., Vol. 35, Springer-Verlag, New York, 1967.

6. N. P. Bhatia, Weak Attractors in Dynamical Systems, Bol. Soc. Mat. Mex. 11 (1966), 56-64.

7. O. Hajek, Absolute Stability of Noncompact Sets, J. Differential Equations 9 (1971), 496-508.

8. O. Hajek, Ordinary and Asymptotic Stability of Noncompact Sets, J. Differential Equations 11 (1972), 49-65.

9. J. R. Isbell, Uniform Spaces, Math. Surveys, No. 12, Amer. Math. Soc., Providence, 1964.

10. S. Willard, General Topology, Addison-Wesley, Reading, 1970.

Chapter 7

NON-UNIFORM SAMPLING AND N-DIMENSION SAMPLING

Kuang-Ho Chen

Department of Mathematics
University of New Orleans
New Orleans, Louisiana

INTRODUCTION

Two sampling theorems concerning n-dimensional uniform sampling and one-dimensional non-uniform sample are proved. In the former case, the constructing function is assumed to have the Fourier transform bandlimited and is uniquely determined. There is no need to assume the existence of the Fourier transform. The Fourier transform derived exists in the distribution sense. The latter case verifies the 1959 Balth van der Pol conjecture for non-uniform sampling, and both imply the one-dimensional uniform sampling. None of the results here apply to the stochastic case.

A survey of the early literature on the problem can be found in Reza [3]. Recently, most work concerns either one-dimensional sampling or statistics, for example, see Masry [2], Todd [4], or Bar-David [1].

The next section studies n-dimensional uniform sampling, and the last section is devoted to non-uniform sampling.

THE N-DIMENSION UNIFORM SAMPLING THEOREM

A distribution F(x) is said to be bandlimited with a cutoff angular frequency of w, where $w = (w_1,\ldots,w_n)$, if F(x) = 0 for $|x_i| > w_i$, for some $i = 1,\ldots,n$. This F is called a w-bandlimited function here for convenience.

THEOREM 1. For a given function f(x) with w-bandlimited Fourier transform, there exists only one f(x) with the representation

$$f(x) = \sum_{m_n=-\infty}^{\infty} \cdots \sum_{m_1=-\infty}^{\infty} f(\pi m_1/w_1, \ldots, \pi m_n/w_n) \cdot \frac{\sin w_1(x_1 - \pi m_1/w_1)}{w_1(x_1 - \pi m_1/w_1)} \cdots \frac{\sin w_n(x_n - \pi m_n/w_n)}{w_n(x_n - \pi m_n/w_n)} \tag{1}$$

Recall, from the well-known Paley-Wiener theorem, that since the Fourier transform F(y) of f(x) is compactly supported, its extension in the complex space C^n is an entire function of finite exponential type. In particular, f(x) is an analytic function. With this fact, this theorem is proved as follows.

Proof. In terms of distributions with support contained in the parallelpipe $P(w) = \{y : |y_i| \leq w_i\}$,

$$f(z) = (F(y), (1/2\pi)^n e^{-izy}) \tag{2}$$

$$F(y) = \sum_{m_n=-\infty}^{\infty} \cdots \sum_{m_1=-\infty}^{\infty} C(m_1,\ldots,m_n)\exp(\pi i(m_1y_1/w_1 + \ldots + m_ny_n/w_n)) \tag{3}$$

$$C(m_1,\ldots,m_n) = (F(y), \exp(\pi i(m_1y_1/w_1+\ldots+m_ny_n/w_n)))/2^nw_1\ldots w_n \tag{4}$$

Substitution of (4) in (3) yields

$$F(y) = \sum_{m_n=-\infty}^{\infty} \cdots \sum_{m_1=-\infty}^{\infty} \pi^n f(\frac{-\pi m_1}{w_1},\ldots,\frac{-\pi m_n}{w_n})/w_1\ldots w_n$$

$$\cdot\ [\exp(\pi i(m_1 y_1/w_1 + \ldots + m_n y_n/w_n))]/w_1\ldots w_n \tag{5}$$

Then, by (2), we have $f(z) = \sum_{m_n=-\infty}^{\infty} \cdots \sum_{m_1=-\infty}^{\infty} f(\frac{-\pi m_1}{w_1},\ldots,\frac{-\pi m_n}{w_n})$

$\cdot\ (\exp(\pi i(m_1 y_1/w_1 + \ldots + m_n y_n/w_n)),\ \exp(-i(z_1 y_1 + \ldots + z_n y_n)))/$
$2^n w_1\ldots w_n$.

The distributions on the right side are just usual integrals over the region P(w). After integrating, interchange each index m_i with $-m_i$, and then equation (1) is obtained. As for uniqueness, let two functions have the expression (1) with same values at $(\pi m_1/w_1,\ldots,\pi m_n/w_n)$. Set f(x) equal to the difference between these two functions. Then, (1) implies that f(x) = 0 for each x in R^n.

Consequently, we have the following commonly used sampling theorem.

COROLLARY 2. For a time-function f(t) with a bandlimited frequency function with a cutoff angular frequency of w,

$$f(t) = \sum_{j=-\infty}^{\infty} f(\frac{j\pi}{w})[\sin w(t-\pi j/w)]/w(t-\pi j/w) \tag{6}$$

THE ONE DIMENSION NON-UNIFORM SAMPLING THEOREM

In 1959, Balth van der Pol [5] made the following statement. Let P(x) be an entire function with simple roots at $\{\ldots, a_1, a_2, \ldots\}$ and f(x) be a bandlimited time function with a cutoff angular frequency π; then, assuming that f(x) and p(x) have no common roots,

$$\frac{f(x)}{p(x)} = \sum_{m=-\infty}^{\infty} \frac{f(a_m)}{p'(a_m)(x-a_m)}\ ,\ \text{(m integers)} \tag{7}$$

Employing standard complex variable techniques, the following non-uniform sampling is proved without assuming f(x) to be a bandlimited time function. The van der Pol conjecture then follows as a consequence.

THEOREM 3. Let P(z) and f(z) be two entire functions with the order of f less than the order of P, or with same order, but the type of f less that of P. Assume further that the roots a_m, (m = 1,2,...) of P(z) are real and simple, and the complement of the roots of f(z) with respect to $\{a_m\}$ is a non-decreasing sequence in absolute values. Then, f(x)/P(x) has the representation (7), which is convergent uniformly on every compact set in the complex plane disjoint from $\{a_m\}$.

Proof. Set q(z) equal to the quotient f(z)/P(z). Then, q(z) is an analytic function with the only possible singularities being $\{a_m\}$, which are just simple poles. Denote by $C(r_m)$ the circle with center at the origin and radius r_m such that for each m, $C(r_m)$ contains in its interior at most 2m points of $\{a_j\}$, and no a_j is on $C(r_m)$. Then, the conditions imposed on f and P about the order and the type, leads to the estimate

$$\limsup_{m \to \infty} \oint_{C(r_m)} |q(z)| \, |dz| = M < +\infty$$

Therefore, all conditions for the well-known Cauchy theorem on partial fraction expansions are satisfied and so, the assertion is proved.

It is remarkable to notice that P(x) can be arbitrarily chosen; only the constructed function f(x) shall have the properties imposed by the theorem. This choice of P(x) is just an extra-freedom on constructing the function f(x). Finally, let us derive the usual one-dimensional uniform sampling theorem (i.e. Corollary 2) from this theorem. Choose P(x) to be sin(wx) and $a_m = m\pi/w$. Then, by the trigonometric identity,

$$\sin(wx) = \sin[w(x - m\pi/w) + m\pi] = \sin w(x - m\pi/w) \cos m\pi$$

we have the representation (6) from (7).

COMMENT

Mr. N. A. Gross produced in his master's degree thesis some results based on the idea presented here. However, his proofs need stronger conditions and his results are much weaker than what appears here.

REFERENCES

1. I. Bar-David, An implicit sampling theorem for bounded band-limited functions, Inform. Contr. 24 (1974), 36-44.

2. E. Masry, The recovery of distorted bandlimited stochastic process, IEEE Trans. Inform. Theory 19 (1973), 398-403.

3. F. M. Reza, An introduction to information theory, McGraw-Hill, New York, 1961.

4. D. E. Todd, Sampled data reconstruction of deterministic bandlimited signals, IEEE Trans. Inform. Theory 19 (1973), 809-810.

5. B. van der Pol, Ann. Computation Lab., Harvard Univ. 29 (1959), 3-25.

Chapter 8

COMPLETE STABILITY AND BOUNDEDNESS OF SOLUTIONS OF A NONLINEAR DIFFERENTIAL EQUATION OF THE FIFTH ORDER

Ethelbert N. Chukwu
Department of Mathematics
Cleveland State University
Cleveland, Ohio

INTRODUCTION

The differential equation considered here is of the form:

$$x^{(5)} + f_1(x,x^{(1)},x^{(2)},x^{(3)},x^{(4)})x^{(4)} + f_2(x^{(2)},x^{(3)})x^{(3)} + f_3(x^{(1)},x^{(2)}) + f_4(x^{(1)}) + f_5(x) = p(t,x,x^{(1)},x^{(2)},x^{(3)},x^{(4)}) \quad (1)$$

in which f_i $(i = 1,2,3,4,5)$ and p are real valued functions which depend at most on the arguments displayed explicitly. It will be assumed that $f_5'(x)$, $f_4'(y)$, $\partial f_3(y,z)/\partial y$, $\partial f_2(z,w)/\partial w$, $\partial f_2(z,w)/\partial z$, $f_1(x,y,z,w,u)$ and $p(t,x,y,z,w,u)$ are continuous for all values of x,y,z,w,u, and t.

The problem of interest here is to investigate conditions under which equation (1) is completely stable (asymptotically stable in the large) or has all solutions bounded. For specialized cases of (1), this problem has received some attention in Chukwu [1] and [2]. This chapter extends these earlier results to equations of the form (1). The stability study is inspired by similar treatments of third order equations by Harrow [4] and Tejumola [9], and fourth order equations by Ezeilo and Tejumola [3], Lalli and Skrapek [5], and Sinha and Hoft [8]. The boundedness result is motivated by a result of Ezeilo and Tejumola [3; Theorem 2]. The nonlinearities in (1) f_i $(i = 1,2,3,4,5)$ satisfy generalized Routh-Hurwitz conditions; the non-Routh-Hurwitz conditions which are imposed are trivial in the constant coefficient situation. These extra conditions have analogues in [3-5, 7-9]. The investigations here rest on a Lyapunov function which is constructed by forming, by a trial method, linear combinations of line integrals. The complete stability result follows from a basic theorem of LaSalle [6].

In what follows we shall deal with the equivalent system

$$x' = y,\ y' = z,\ z' = w,\ w' = u$$

$$u' = -f_1(x,y,z,w,u)u - f_2(z,w)w - f_3(y,z)$$
$$- f_4(y) - f_5(x) + p(t,x,y,z,w,u)$$

obtained from (1).

NOTATION. In what follows the letters D and D_i represent positive finite constants whose magnitudes depend only on the constants which appear in the statements of Theorem 1 and 2 below as well as the functions f_i $(i = 1,2,3,4,5)$. They are independent of any particular solutions chosen. While the D_i retain their identities in each place of occurence the values of D may vary.

STATEMENTS OF RESULTS

THEOREM 1. Consider (2) with $p(t,x,y,z,w,u) \equiv 0$ and suppose that the following conditions hold.

H_1. There are positive constants a,b,c,d,e such that $a > 0$, $ab - c > 0$, $(ab-c)c - (ad-e)a > 0$, $e > 0$, and $\Delta \equiv (dc-be)(ab-c) - (ad-e)^2 > 0$,

$\Delta_1 \equiv (dc-be)(ab-c)/(ad-e) - (af_4'(y)-e) > 2b\varepsilon$ and
$\Delta_2 \equiv (dc-be)/(ad-e) - \gamma(ad-e)/d(ab-c) - \varepsilon/a > 0$,
where $\gamma = f_4(y)/y$ if $y \neq 0$ and $\gamma = f_4'(0)$ if $y = 0$,
and ε is sufficiently small positive constant.

H_2. Let $f_3(y,0) = 0 = f_4(0) = f_5(0)$ for all y.

H_3. Assume that $f_1(x,y,z,w,u) \geq a+2k$, for some $k > 0$, $f_2(z,w) \geq b$, $f_3(y,z)/z \geq c$ for $z \neq 0$, $f_4(y)/y \geq d$ for $y \neq 0$, and $f_5'(x) \leq e$.

H_4. Let $f_5(x)$ sgn $x > 0$ if $x \neq 0$, and $F_5(x) \equiv \int_0^x f_5(s)ds \to \infty$ as $|x| \to \infty$.

H_5. Suppose that $f_1(x,y,z,w,u) - a < \min [\varepsilon/32a^2, \varepsilon d/16\delta^2, \Delta_1(\frac{ad-e}{ab-c})^2/64d^2]$, $f_2(z,w) - b < \varepsilon ad/2\delta^2$, $(f_3(y,z)/z - c)^2 < \min [k\Delta_1/16, \varepsilon d\Delta_1/64\delta^2]$ if $z \neq 0$, $(d - f_4'(y))^2 < \varepsilon\Delta_1/128$, and $e - f_5'(x) < \varepsilon d(ab-c)/8(ad-e)$, where $\delta = e(ab-c)/(ad-e) + \varepsilon$.

H_6. Let $z\partial f_2(z,w)/\partial w \geq 0$, $w\partial f_2(z,w)/\partial z < \varepsilon/2$,

$$1/z \int_0^z (\partial f_3(y,x)/\partial y)ds < \Delta_1/4a \text{ if } z \neq 0, \text{ and}$$

$$f_4'(y) - f_4(y)/y < e\Delta/d^2(ab-c) \text{ if } y \neq 0.$$

Then every solution $(x(t),y(t),z(t),w(t),u(t))$ of (2) satisfies $x^2(t) + y^2(t) + z^2(t) + w^2(t) + u^2(t) \to 0$ as $t \to \infty$.

The usual Routh Hurwitz restrictions in H_1 imply that $a > 0$, $b > 0$, $c > 0$, $d > 0$, $e > 0$, $ad-e > 0$, $dc-be > 0$. The special case $f_1 = a$, $f_2(z,w)w = g_2(w)$, $f_3(y,z) = c$ was studied in [2], while the case $f_2(z,w) \equiv b$, $f_3(y,z) = g_3(z)$ was studied in [1]. The conditions H_6 are comparable to analogous criteria in [9; Theorem 1(iii)], [3; Theorem 1(iii)], and [8; Theorem 1(ii)].

THEOREM 2. Suppose that in (2), $f_3(y,0) = f_4(0) = 0$, and conditions H_1, H_3, H_5 and H_6 hold. Furthermore, assume that (i) $f_5(x)$ sgn $x > 0$ for $|x| \geq 1$, and (ii) the function $p(t,x,y,z,w,u)$ satisfies

$$|p(t,x,y,z,w,u)| \leq \{A+|y|+|z|+|w|+|u|\}\psi(t) \qquad (3)$$

where A is a constant and $\psi(t) \geq 0$ is a continuous function of t. Then for any finite x_0, y_0, z_0, w_0, u_0, there exist constants $K_i = K_i(x_0, y_0, z_0, w_0, u_0)$ $i = 0,1$ and a constant $\lambda > 0$ whose magnitude is independent of x_0, y_0, z_0, w_0, u_0 such that any solution $x(t), y(t), z(t), w(t), u(t)$ of (2) determined by

$$x(0) = x_0,\ y(0) = y_0,\ z(0) = z_0,\ w(0) = w_0,\ u(0) = u_0 \quad (4)$$

satisfies, for all $t \geq 0$,

$$y^2(t)+z^2(t)+w^2(t)+u^2(t) \leq K_0\{1+\chi^{-1}(t)[1+\int_0^t \psi(s)\chi(s)ds]\}$$

$$F_5(x(t)) = \int_0^{x(t)} f_5(s)ds \leq K_1\{1+\chi^{-1}(t)[1+\int_0^t \psi(s)\chi(s)ds]\}$$

where $\chi(t) = \exp(-\lambda\int_0^t \psi(s)ds)$.

The following corollary is an immediate consequence of Theorem 2.

COROLLARY 3. Suppose, in addition to the conditions of Theorem 2, that $F_5(x) \to +\infty$ as $|x| \to \infty$, and $\int_0^\infty \psi(t)dt < \infty$. Then there exists a constant $K_2 = K_2(x_0, y_0, z_0, w_0, u_0)$ such that the unique solution $(x(t), y(t), z(t), w(t), u(t))$ of (2) determined by (4) satisfies $|x(t)| \leq K_2$, $|y(t)| \leq K_2$, $|z(t)| \leq K_2$, $|w(t)| \leq K_2$, and $|u(t)| \leq K_2$, for all $t \geq 0$.

SOME PRELIMINARY LEMMAS

The proof of the results above rest on the following Lyapunov function

$$\begin{aligned} 2V = {} & u^2 + 2auw + 2d(ab-c)uz/(ad-e) + 2\delta yu \\ & + 2\int_0^w sf_2(z,s)ds + [a^2-d(ab-c)/(ad-e)]w^2 \\ & + 2[c+ad(ab-c)/(ad-e)-\delta]wz + 2a\delta wy + 2wf_4(y) \end{aligned}$$

$$+ 2wf_5(x) + 2a\int_0^z f_3(y,s)ds - (d+a\delta)z^2 + 2\delta byz$$

$$+ 2d(ab-c)\int_0^z sf_2(s,0)ds/(ad-e) + 2azf_4(y)$$

$$- 2ezy + 2azf_5(x) + 2d(ab-c)\int_0^y f_4(s)ds/(ad-e)$$

$$+ (\delta c-ea)y^2 + 2d(ab-c)yf_5(x)/(ad-e) + 2\delta\int_0^x f_5(s)ds$$

where δ is defined by $\delta = e(ab-c)/(ad-e) + \varepsilon$.

The required properties of V will be stated in the next two lemmas; their proofs will not be given. The full force of Lemma 4 may not be required. The well-known theorem of LaSalle [6] which we shall apply, requires that $V \to \infty$ as $x^2+y^2+z^2+w^2+u^2 \to \infty$, but there seems to be no direct or simpler way of showing this other than by the way of Lemma 4.

LEMMA 4. Assume that all the conditions of Theorem 1 hold. Then there are positive constants D_i $(i = 1,2,3,4,5)$ such that for all x,y,z,w, and u,

$$2V \geq D_1F_5(x) + D_2y^2 + D_3z^2 + D_4w^2 + D_5u^2$$

provided that ε is sufficiently small.

LEMMA 5. Assume that all the conditions of Theorem 1 hold. Then there exist positive constants D_i $(i = 6,7,8)$ such that if $(x(t),y(t),z(t),w(t),u(t))$ is a solution of (2) with $p(t,x,y,z,w,u) \equiv 0$ then

$$V' \leq -ku^2 - (D_6y^2 + D_7z^2 + D_8w^2)$$

PROOFS OF THE THEOREMS

PROOF OF THEOREM 1. From Lemma 4 and Lemma 5 we have

$$V(x,y,z,w,u) > 0 \text{ for } x^2+y^2+z^2+w^2+u^2 > 0 \tag{5}$$

$$V \to +\infty \text{ as } x^2+y^2+z^2+w^2+u^2 \to \infty \tag{6}$$

and

$$V' \leq -D(y^2+z^2+w^2+u^2) \tag{7}$$

where $D > 0$ is a constant. It follows from the system

$$x' = y,\ y' = z,\ z' = w,\ w' = u \tag{8}$$

$$u' = -f_1(x,y,z,w,u)u-f_2(z.w)-f_3(y\ z)-f_4(y)-f_5(x)$$

hypothesis H_2 and (7) that the set E of all solutions of (8) such that $V' = 0$ consists of just the origin. Thus the largest invariant set M in E is the origin. It also follows from (7) that $V(x(t),y(t),z(t),w(t),u(t) \leq V(x(0),y(0),z(0),w(0),u(0))$ for all $t \geq 0$, so that by (6) all solutions of (8) are bounded for all $t \geq 0$. It is now clear that all the conditions of Theorem 3 of LaSalle [6] are satisfied. Since M is the origin. complete stability (asymptotic stability in the large) follows at once.

PROOF OF THEOREM 2. The proof is analogous to the proof of Theorem 2 in Chukwu [2].

It is no longer true that the estimate for 2V in Lemma 4 holds valid under the restriction (i) of Theorem 2. However, it is rather simple to verify that V, at least, satisfies

$$2V \geq D_1F_5(x) + D_2y^2 + D_3z^2 + D_4w^2 + D_5u^2 - 2D_9$$

for some D_9; it follows that

$$V \geq D_{10}(y^2+z^2+w^2+u^2) + D_{11}F_5(x) - D_9$$

for sufficiently small D_{10}. Also, since $f_5(x)$ sgn $x > 0$ for $|x| > 1$, and $f_5(x)$ continuous, there exists a D_{12} such that $F_5(x) \geq -D_{12}$ for all x. Therefore

$$V \geq D_{10}(y^2+z^2+w^2+u^2) - D_{13} \tag{9}$$

where $D_{13} = D_{11}D_{12} + D_9$. Now let $(x(t),y(t),z(t),w(t),u(t))$ be the solution of (2) satisfying the initial conditions (4) and set $V(t) = V(x(t),y(t),z(t),w(t),u(t))$. It follows, just as in Lemma 5, that

$$V' \le -D(y^2+z^2+w^2)+[a+aw+d(ab-c)z/(ad-e)+\delta y]p(t,x,y,z,w,u)$$

so that

$$V' \le D_{14}(|y|+|z|+|w|+|u|)|p(t,x,y,z,w,u)|$$

Thus, by (3) and the obvious inequalities $|y| \le 1+y^2$, $|z| \le 1+z^2$, $|w| \le 1+w^2$, $|u| \le 1+u^2$, and $(|y|+|z|+|w|+|u|)^2 \le 4(y^2+z^2+w^2+u^2)$, we have $V' \le D_{14}[4A+(A+4)(y^2+z^2+w^2+u^2)]\psi(t)$ so

$$V' - D_{15}V(t)\psi(t) \le D_{16}\psi(t)$$

On multiplying both sides by $\chi(t) = \exp(-D_{15}\int_0^t \psi(s)ds)$ and integrating, we obtain

$$V(t)\chi(t) \le V(0)+D_{16}\int_0^t \psi(s)\chi(s)ds$$

for $t \ge 0$, and on dividing both sides by $\chi(t)$ we have

$$V(t) \le \chi^{-1}(t)[V(0)+D_{16}\int_0^t \psi(s)\chi(s)ds] \tag{10}$$

where $V(0) = V(x_0,y_0,z_0,w_0,u_0)$. This together with (9) shows that

$$y^2+z^2+w^2+u^2 \le D_{10}{}^{-1}[\chi^{-1}(t)\{V(0)+D_{16}\int_0^t \psi(s)\chi(s)ds\}+D_{13}]$$

The other conclusion follows from (10) and the fact that $D_{11}F_5(x) \le V+D_9$. The proof is now complete.

ACKNOWLEDGEMENT

The research reported here, done at the Centre for Dynamical Systems at Brown University (Providence, Rhode Island), was supported in part by a 1974 Cleveland State University Research Initiation Award.

REFERENCES

1. E. N. Chukwu, On the Boundedness and Stability Properties of Solutions of Some Differential Equations of the Fifth Order, Ann. Mat. Pura Appl. (4) 106 (1975), 245-258.

2. E. N. Chukwu, On the Boundedness and the Stability of Solutions of Some Differential Equations of the Fifth Order, SIAM J. Math. Anal. 7 (1976), 176-194.

3. J. O. C. Ezeilo and H. O. Tejumola, On the Boundedness and the Stability Properties of Solutions of Certain Fourth Order Differential Equations, Ann. Mat. Pura Appl. (4) 95 (1973), 132-145.

4. M. Harrow, Further Results for the Solutions of Certain Third Order Differential Equations, J. London Math. Soc. 43 (1968), 587-592.

5. B. S. Lalli and W. A. Skrapek, On the Boundedness and Stability of Some Differential Equations of the Fourth Order, SIAM J. Math. Anal. 2 (1971), 221-225.

6. J. P. LaSalle, The Extent of Asymptotic Stability Proc. Nat. Acad. Sci., U.S.A. 46 (1960). 363-365.

7. A. S. C. Sinha and Y. Hari, On the Boundedness of Solutions of Some Nonautonomous Differential Equations of the Fourth Order, Inter. J. Control 15 (1972), 717-724.

8. A. S. C. Sinha and R. G. Hoft, Stability of a Nonautonomous Differential Equation of the Fourth Order, SIAM J. Control 9 (1971), 8-14.

9. H. O. Tejumola, A Note on the Boundedness and the Stability of Solutions of Certain Third-Order Differential Equations, Ann. Mat. Pura Appl. (4) 92 (1972), 65-75.

Chapter 9

STABILITY TYPE PROPERTIES FOR SECOND ORDER NONLINEAR DIFFERENTIAL EQUATIONS

JOHN R. GRAEF AND PAUL W. SPIKES

Department of Mathematics
Mississippi State University
Mississippi State, Mississippi

INTRODUCTION

Stability type properties of second order nonlinear differential equations of the form

$$(a(t)x')' + h(t,x,x') + q(t)f(x)g(x') = e(t,x,x') \qquad (*)$$

have been studied by a number of authors. As examples we cite the recent work of Baker [1], Graef and Spikes [3-5], Grimmer et. al. [2,6], Hammett [7], Kartsatos [8], Londen [9], and Wong [10,11]. In this chapter we discuss the boundedness and convergence to zero of solutions of (*) without making the usual assumption that the perturbation term $e(t,x,x')$ is small. Other conditions often required by other authors have also been relaxed.

The first three theorems concern the boundedness of all solutions and the convergence to zero of the nonoscillatory and Z-type solutions. Theorems 5 and 6 give sufficient conditions for the oscillatory and Z-type solutions to converge to zero, and are extensions of results of Wong [10,11]. Complete details of the results in this paper will appear in [4].

STABILITY PROPERTIES

Consider the equation

$$(a(t)x')' + h(t,x,x') + q(t)f(x)g(x') = e(t,x,x') \tag{1}$$

where $a,q : [t_0,\infty) \to R$, $f,g : R \to R$, and $h,e : [t_0,\infty) \times R^2 \to R$ are continuous, $a(t) > 0$, $q(t) > 0$, and $g(x') > 0$. It will be convenient to write equation (1) as the system

$$\begin{aligned} x' &= y \\ y' &= (-a'(t)y-h(t,x,y)-q(t)f(x)g(y)+e(t,x,y))/a(t) \end{aligned} \tag{2}$$

Let $q'(t)_+ = \max\{q'(t),0\}$ and $q'(t)_- = \max\{-q'(t),0\}$ so that $q'(t) = q'(t)_+ - q'(t)_-$. Define $F(x) = \int_0^x f(s)ds$, $G(y) = \int_0^y [s/g(s)]ds$ and assume that there is a continuous function $r : [t_0,\infty) \to R$ such that

$$|e(t,x,y)| \le r(t) \tag{3}$$

and

$$h(t,x,y)y \ge 0 \tag{4}$$

We will use the same classification of solutions that was used in [3-5]. That is, a solution $x(t)$ of (1) will be called nonoscillatory if there exists $t_1 \ge t_0$ such that $x(t) \ne 0$ for $t \ge t_1$; the solution will be called oscillatory if for any given $t_1 \ge t_0$ there exist t_2 and t_3 satisfying $t_1 < t_2 < t_3$, $x(t_2) > 0$, and $x(t_3) < 0$; and it will be called a Z-type solution if it has arbitrarily large zeros but is ultimately nonnegative or nonpositive. The following additional assumptions are needed in order to show that nonoscillatory solutions of (1) converge to zero. Assume that:

(i) $xf(x) > 0$ if $x \ne 0$ and $f(x)$ is bounded away from zero if x is bounded away from zero

(ii) condition (3) holds and $r(t)/q(t) \to 0$ as $t \to \infty$

(iii) if x is bounded, then there exists a continuous function k and $t_1 \geq t_o$ such that $|h(t,x,y)| \leq k(t)g(y)$ for (t,x,y) in $[t_1,\infty) \times R^2$ and $k(t)/q(t) \to 0$ as $t \to \infty$

(iv) $g(y) \geq c > 0$, $\int_{t_o}^{\infty} q(s)ds = \infty$, and $\int_{t_o}^{\infty} [1/a(s)]ds = \infty$

LEMMA 1. If (i) - (iv) hold and $x(t)$ is a bounded non-oscillatory solution of (1), then $\liminf_{t\to\infty} |x(t)| = 0$.

To see that condition (iii) in Lemma 1 is essential, observe that both the equations

$$x'' + tx' + x/t = 1/t^2 + 2/t^3, \quad t > 0$$

$$x'' + tx' + x[1+(x')^2]/t = (t^4+2t^3+t+1)/t^6, \quad t > 0$$

have the nonoscillatory solution $x(t) = (1 + t)/t$. In the first equation we do not have $|h(t,x,x')| \leq k(t)g(x')$ and in the second equation we do not have $k(t)/q(t) \to 0$ as $t \to \infty$.

The proof of Lemma 1 as well as the proofs of the following three theorems will appear in [4]. We will also make use of the following notation.

CONDITION W. If $x(t)$ is a nonoscillatory or Z-type solution of (1), then $\lim_{t\to\infty} x(t) = 0$.

THEOREM 2. Suppose that conditions (3) and (4) hold,

$$F(x) \to \infty \text{ as } |x| \to \infty \tag{5}$$

$$\int_{t_o}^{\infty} [a'(s)_-/a(s)]ds < \infty \text{ and } a(t) \leq a_1 \tag{6}$$

$$\int_{t_o}^{\infty} [q'(s)_-/q(s)]ds < \infty \tag{7}$$

$$\int_{t_o}^{\infty} [r(s)/q(s)]ds < \infty \tag{8}$$

and there is a positive constant N such that

$$y^2/g(y) \leq N \tag{9}$$

Then all solutions of (1) are bounded. If, in addition (i) - (iv) hold, then Condition W holds.

THEOREM 3. If (3) - (5) and (7) - (8) hold, there is a positive constant L such that $|y|/g(y) \leq L$, and

$$a'(t) \geq 0 \text{ and } a(t) \leq a_1 \tag{10}$$

then all solutions of (1) are bounded. Under the additional assumptions (i) - (iv), Condition W holds.

THEOREM 4. Suppose conditions (3) - (7) hold, $g(y) \geq c > 0$, there are positive constants M and k such that

$$y^2/g(y) \leq MG(y) \text{ for } |y| \geq k \tag{11}$$

and

$$\int_{t_o}^{\infty} [r(s)/(q(s))^{1/2}]ds < \infty \tag{12}$$

Then all solutions of (1) are bounded. Moreover, if (i) - (iv) hold, then Condition W is satisfied.

Examples showing the relationship between Theorems 1-3 and some recent results of Hammett [7], Grimmer [6] and Londen [9] can be found in [4].

The next two theorems give sufficient conditions for the oscillatory and Z-type solutions of (1) to converge to zero. They are extensions of some results of Wong [10,11]. We shall assume that

$$xf(x) > 0 \text{ if } x \neq 0 \tag{13}$$

$$q(t) \to \infty \text{ as } t \to \infty \tag{14}$$

$$0 < c < g(y) < C \text{ and } |a'(t)| \leq a_2 \tag{15}$$

$$H(t) = r(t)/q(t) \to 0 \text{ as } t \to \infty \tag{16}$$

$$xf(x) \geq dF(x) \tag{17}$$

for some positive constant d, and there is a continuous function $k : [t_o,\infty) \to R$ such that

$$|h(t,x,y)| \le k(t) \text{ and } k(t)/q(t) \to 0 \text{ as } t \to \infty \tag{18}$$

THEOREM 5. Suppose conditions (3) - (7) and (12) - (18) hold. If

$$\int_{t_o}^{t} |(q^{-1}(s))'''|ds = o(\ln q(t)), \quad t \to \infty$$

then every oscillatory or Z-type solution x(t) of (1) satisfies $\lim_{t\to\infty} x(t) = 0$.

THEOREM 6. Let conditions (3) - (5), (10), and (12) - (18) hold. If for every w with 1/2 < w < 1 we have

$$\int_{t_o}^{\infty} [q'(s)_-/q^w(s)]ds < \infty \tag{19}$$

and

$$\int_{t_o}^{t} |(q^{-w}(s))'''|ds = o(q^{1-w}(t)), \quad t \to \infty \tag{20}$$

then every oscillatory or Z-type solution x(t) of (1) satisfies $\lim_{t\to\infty} x(t) = 0$.

We will outline a proof of this theorem. Complete details of the proofs of both Theorems 5 and 6 will appear in [4].

Let x(t) be an oscillatory or Z-type solution of (1). First note that conditions (14) and (19) imply that (7) holds so by Theorem 4 x(t) is bounded, say $|x(t)| \le B$. With no loss in generality we may assume that C > dc. Let N = (4C - dc)/2dc and w = Ndc/(2C + dc); then 1/2 < w < 1. For $t \ge z \ge t_o$ define

$$V_z(x,y,t) = F(x)/a(t) + G(y)/q(t)$$
$$+ \int_z^t [h(s,x(s),y(s))y(s)/g(y(s))q(s)a(s)]ds$$
$$- \int_z^t [e(s,x(s),y(s))y(s)/g(y(s))q(s)a(s)]ds$$

It can be shown that $V_z(t)$ approaches a finite limit as $t \to \infty$, so there exists $z \geq t_o$ such that

$$\int_z^\infty [h(s,x(s),y(s))y(s)/g(y(s))q(s)a(s)]ds < N\varepsilon(1-w)15w$$

$$\int_z^\infty |e(s,x(s),y(s))y(s)/g(y(s))q(s)a(s)|ds$$
$$< \min\{N\varepsilon(1-w)/15w,\ \varepsilon/8\}$$

and

$$B[r(t) + k(t)]/q(t)a(t) < Ndc\varepsilon(1-w)/15w$$

for $t \geq z$.

Let $T(t) = 1/q^w(t)$ and $K(t) = NdcT(t)q(t)V_z(t) + T''(t)x^2/2 - T'(t)xy$. Then $K'(t) \leq T'''(t)x^2/2 + c_1q'(t)/q^b(t) + c_2q'(t)_-/q^w(t) + Dq'(t)_+V_z(t)/q^w(t) - Dq'(t)_-V_z(t)/q^w(t) + dc|q'(t)|[2N\varepsilon(1-w)/15]q^w(t) + Ndc\varepsilon(1-w)|q'(t)|/15q^w(t)$. Since $V_z(t)$ converges, suppose that $\lim_{t\to\infty} V_z(t) = L \geq 3\varepsilon/4$. Now $x(t)$ is an oscillatory or Z-type solution so we let $\{t_n\}$ be an increasing sequence of zeros of $y(t)$ such that $t_n \to \infty$ as $n \to \infty$ and $4L/5 \leq V_z(t) \leq 6L/5$ for $t \geq t_1$. Hence $K'(t) \leq |T'''(t)|B^2/2 + c_1q'(t)/q^b(t) + c_2q'(t)_-/q^w(t) + 6DLq'(t)/5q^w(t) + 4NdcL(1-w)q'(t)/15q^w(t)$. Integrating from t_1 to t_n we have

$$K(t_n) \leq (B^2/2)\int_{t_1}^{t_n}|T'''(s)|ds + c_1q^{1-b}(t_n)/(1 - b)$$
$$+ c_3\int_{t_1}^{t_n}[q'(s)_-/q^w(s)]ds + 6DLq^{1-w}(t_n)/5(1 - w)$$
$$+ 4NdcLq^{1-w}(t_n)/15 + c_4$$

Now $K(t_n) = NdcT(t_n)q(t_n)V_z(t_n) + T''(t_n)x^2(t_n)/2 \geq 4NdcLq^{1-w}(t_n)/5 + T''(t_n)x^2(t_n)/2$, and since $b > 1$ and $[1 - w(1 + 1/N)]/(1 - w) = 1/3$, we have

$$2NdcLq^{1-w}(t_n)/15 \leq B^2\int_{t_1}^{t_n}|T'''(s)|ds + c_5$$

which is impossible in view of (14) and (20). Therefore $\lim_{t\to\infty} V_z(t) = L < 3\varepsilon/4$. Hence there exists $T_1 \geq z$ such that $V_z(t) < 7\varepsilon/8$ for $t \geq T_1$ so

$$\begin{aligned} F(x(t))/a(t) & \\ &< 7\varepsilon/8 + \int_{T_1}^{t} |e(s,x(s),y(s))y(s)/g(y(s))q(s)a(s)|ds \\ &< \varepsilon \end{aligned}$$

for $t \geq T_1$. Since $a(t)$ is bounded from above, we have that $F(x(t)) \to 0$ as $t \to \infty$ which implies that $x(t) \to 0$ as $t \to \infty$ completing the proof of the theorem.

By combining various theorems we could obtain results which would guarantee that all solutions of (1) tend to zero as $t \to \infty$. One such example is the following.

THEOREM 7. If conditions (3) - (7), (10) - (20), and (i) - (iv) hold, then every solution $x(t)$ of (1) satisfies $\lim_{t\to\infty} x(t) = 0$.

ACKNOWLEDGEMENT

This research was supported by the Mississippi State University Biological and Physical Sciences Research Institute.

REFERENCES

1. J. W. Baker, Stability properties of a second order damped and forced nonlinear differential equation, SIAM J. Appl. Math. 27 (1974), 159-166.
2. T. A. Burton and R. C. Grimmer, Stability properties of $(r(t)u')' + a(t)f(u)g(u') = 0$, Monatsh. Math. 74 (1970), 211-222.
3. J. R. Graef and P. W. Spikes, Asymptotic behavior of solutions of a second order nonlinear differential equation, J. Differential Equations 17 (1975), 461-476.
4. J. R. Graef and P. W. Spikes, Boundedness and convergence to zero of solutions of a forced second order nonlinear differential equation, J. Math. Anal. Appl., to appear.

5. J. R. Graef and P. W. Spikes, Continuability, boundedness, and asymptotic behavior of solutions of $x'' + q(t)f(x) = r(t)$, Ann. Mat. Pura Appl. (4) 101 (1974), 307-320.

6. R. Grimmer, On nonoscillatory solutions of a nonlinear differential equation, Proc. Amer. Math. Soc. 34 (1972), 118-120.

7. M. E. Hammett, Nonoscillation properties of a nonlinear differential equation, Proc. Amer. Math. Soc. 30 (1971), 92-96.

8. A. G. Kartsatos, On the maintenance of oscillations of nth order equations under the effect of a small forcing term, J. Differential Equations 10 (1971), 355-363.

9. S. Londen, Some nonoscillation theorems for a second order nonlinear differential equation, SIAM J. Math. Anal. 4 (1973), 460-465.

10. J. S. W. Wong, Remarks on stability conditions for the differential equation $x'' + a(t)f(x) = 0$, J. Austral. Math. Soc. 9 (1969), 496-502.

11. J. S. W. Wong, Some stability conditions for $x'' + a(t)x^{2n-1} = 0$, SIAM J. Appl. Math. 15 (1967), 889-892.

Chapter 10

PRICE STABILITY IN UNIONS OF MARKETS

CHARLES R. JOHNSON*

Department of Economics
Institute for Physical Science and Technology
University of Maryland
College Park, Maryland

CLASSICAL EQUILIBRIUM MODELS

We consider markets in the set of commodities $C_1, \dots, C_n$, $n \geq 2$. Let $p_i \geq 0$ be a variable indicating the price of C_i, $i = 1, \dots, n$, and let $p = (p_1, \dots, p_n)$ denote the vector of prevailing prices. Given the prices p, each potential market buyer of C_i is assumed to demand a certain quantity of C_i while each potential supplier is assumed willing to supply some quantity of C_i. By simple summation, the total market demand for C_i at prices p is obtained and, likewise, the total market supply of C_i at prices p. The former yields the demand function $d_i(p)$, and the latter yields the market supply function $s_i(p)$ for C_i. We may then define for each $i = 1, \dots, n$,

$$f_i(p) \equiv d_i(p) - s_i(p)$$

the market "excess demand function" for C_i at the prices p. The traditional assumption is that if $f_i(p) > 0$ then p_i should

*Other affiliation: Applied Mathematics Division, National Bureau of Standards, Washington, D. C.

tend to rise over time, while, if $f_i(p) < 0$, then p_i should tend to fall over time [14]. This notion first suggests the very simple model

$$\frac{dp_i}{dt} = f_i(p), \quad i = 1,\dots,n \tag{1}$$

The price vector $\hat{p}$ is then naturally defined to be an equilibrium if

$$f_i(\hat{p}) = 0, \quad i = 1,\dots,n \tag{2}$$

In order to allow for more general rates of adjustment, the slightly more complicated model

$$\frac{dp_i}{dt} = H(f_i(p)), \quad i = 1,\dots,n \tag{3}$$

may be advanced where H is a differentiable function satisfying $H(0) = 0$ and $H'(0) > 0$ [1]. The notion of equilibrium is then still as indicated by (2). We do not address the issue here, but the existence of equilibrium for market models such as (1) (and, thus, (3)) is known under quite general circumstances [2].

STABILITY ANALYSIS OF EQUILIBRIUM

An equilibrium price vector $\hat{p}$ is said to be stable (locally asymptotically stable) if, after any "small", one-time perturbation, the system (1) (or (3)) tends to return to $\hat{p}$ over time. If all partials of the excess demand functions f_i, $i = 1,\dots,n$, exist, then we may define the n-by-n Jacobian

$$A = \left(\frac{\partial f_i}{\partial p_i} (\hat{p}) \right) \tag{4}$$

of the system (1), which we shall call the Jacobian of the excess demand functions at $\hat{p}$. It is then well-known that $\hat{p}$ is stable if the real part of each eigenvalue of A is negative [5]. Such a matrix is thus also called stable. Since the Jacobian of the system (3) is simply $H'(0)A$ and $H'(0) > 0$, the stability of $\hat{p}$ in (3) is equivalent to the stability of $\hat{p}$ in (1).

For an n-by-n matrix B, we denote the "real part" (or "Hermitian part") of B by $\text{Re}(B) = (B+B^*)/2$. The classical result of Lyapunov [12] then characterizes the stability of A (and thus $\hat{p}$) in the following way: "A is stable if and only if there is a positive definite Hermitian matrix G such that

$$-\mathrm{Re}(GA) \tag{5}$$

is positive definite". The set of all positive definite Hermitian G for which (5) is positive definite is easily verified to form a cone, and we shall denote this cone by L(A).

Remark 1. Thus A is stable if and only if $L(A) \neq \emptyset$.

A condition sufficient, but not necessary, for the stability of $A = (a_{ij})$ may be deduced from Gersgorin's theorem [15]. If $a_{ii} < 0$ and DA is diagonally dominant of its columns for some positive diagonal matrix D, then all Gersgorin column discs for DAD^{-1}, and thus all eigenvalues of A, lie in the open left half-plane so that A is stable. Denote by D(A) the set all positive diagonal matrices D such that $a_{ii} < 0$ and DA is column diagonally dominant, and then D(A) is also a cone.

Remark 2. If $D(A) \neq \emptyset$, then A is stable.

Global asymptotic stability of systems such as (3) is less well understood and is only known in certain special cases [2,13]. Most notable among these is the case of "gross substitutes", that is, when the Jacobian of the excess demand functions is an M-matrix at all price vectors [3]. In this case $D(A) \neq \emptyset$ for all p and it is reasonable to ask if this condition is sufficient for global asymptotic stability in other than the gross substitute case. As of this writing the answer is not known [13].

THE PROBLEM OF D-STABILITY

The model (3) may be further complicated by supposing, realistically, that the adjustment rates differ from commodity to commodity within the market (i.e. from i-th market to the j-th market, $i \neq j$, within the large market). This suggests the model

$$\frac{dp_i}{dt} = H_i(f_i(p)) \tag{6}$$

where H_i is differentiable and satisfies $H_i(0) = 0$ and $H_i'(0) > 0$, $i = 1,...,n$. The H_i's might be thought of as commodity-specific adjustment functions. Stability analysis of this system is complicated by the fact that the adjustment functions are taken to be unknown [1], other than the above requirements, and this renders the analysis intrinsically different from that of (3). The Jacobian of (6) at an equilibrium $\hat{p}$ is simply DA where A is the Jacobian of the excess demand functions and D is the positive diagonal matrix whose i-th diagonal entry is $H_i'(0)$. For $\hat{p}$ to be a stable equilibrium regardless of the adjustment rates, DA must be stable for all positive diagonal matrices D. Such a matrix A is called D-stable. No effective characterization of the D-stable matrices is known despite a good deal of research [6], and one would be most welcome. However, because of the main result of [7], it is clear that this problem is difficult. It is not difficult to demonstrate that A is D-stable if $D(A) \neq \emptyset$ and also, more generally, that A is D-stable if L(A) contains a diagonal matrix; however, neither of these conditions is necessary. For D(A) to be nonempty, effective necessary and sufficient conditions are well-known in terms of the positivity of the leading principal minors of a matrix derived from A [6]. On the other hand no precise method of determining when L(A) contains a diagonal matrix seems to be known, although some sufficient and other necessary conditions can be formulated [8]. Here there is room for interesting further research on both a theoretical and a numerical level.

Two economically interesting variations on the D-stability question should also be mentioned because they too merit further study. If the adjustment rates comprising D, while still unknown, are known to fall within certain ranges, then the class of matrices A, stable under multiplication by only the allowable D's, would be of interest. Such a matrix A might be called conditionally D-stable, and, generally speaking, the greater the knowledge of the adjustment rates, the larger would be the class of conditionally D-stable A's.

If, further, the adjustment rates were not only known but were regarded as controllable policy variables, then, if achievement of stability were the goal, the existence of a positive diagonal matrix D such that DA is stable would be at issue. Such matrices A might be called D-stabilizable, and an effective characterization for them is also lacking. The Fisher-Fuller conditions are sufficient (an elegant and simple proof is given in [4]) but not necessary. The consummate mathematical question one is led to is to characterize, for a given matrix A, the set of all positive diagonal matrices D such that DA is stable. This would, of course, contain the questions of D-stability, conditional D-stability, and D-stabilizability.

UNIONS OF TWO MARKETS

We next consider the case of two entirely separate markets, both in the same set of commodities $C_1,\ldots,C_n$. Then, imagine the two markets to be instantaneously joined into one (e.g. upon the removal of a high tariff barrier or after the revolutionary lowering of a significant transportation cost etc.). The market supply and demand functions, and thus the excess demand function, for C_i in the new market will then just be the sum of those from the two old markets. Because of the linearity of the derivative, the new Jacobian of the excess demand functions will be the sum of the two prior Jacobians. If a given price vector $\hat{p}$ is an equilibrium in both of the old markets, then $\hat{p}$ will again be an equilibrium in the surviving market. However, even if $\hat{p}$ is a stable equilibrium in both markets, it is not necessarily stable in the union since, as can easily be shown by example, the sum of two stable matrices is not always stable.

In the following we assume the very simplest case: that our markets obey the model (1) and that $\hat{p}$ is an equilibrium price vector common to and stable in each of the original markets. We denote by A the Jacobian of the excess demand functions of the first market and by B the Jacobian of the second. Equivalent to the question of the stability of $\hat{p}$ in the union of the two markets, then, is the mathematical question: "given two stable matrices A and B, when is A+B stable?" In economic terms, what must the structure of the two markets have in common for the union to be stable?

SUFFICIENT CONDITIONS FOR STABILITY OF SUMS

It is a straightforward computation to show that

$$D(A) \cap D(B) \subseteq D(A+B)$$

and that

$$L(A) \cap L(B) \subseteq L(A+B)$$

From these observations, two sufficient conditions for the stability of A+B follow.

THEOREM 1. If A and B are stable matrices and $\emptyset \neq D(A) \cap D(B)$ then A+B is stable.

THEOREM 2. If A and B are stable matrices and $\emptyset \neq L(A) \cap L(B)$ then A+B is stable.

Theorem 2 says, for example, that if the two Jacobians have a common Lyapunov solution, then the common equilibrium is stable in the union of the two markets. Unfortunately, the converse of neither theorem is valid as may be shown via 2-by-2 examples. It is still an interesting question to determine for the stable pair A and B nontrivial necessary and sufficient conditions which insure that A+B is stable. In the next two sections we give a characterization for a slight variation on this problem.

A "LOCAL" LYAPUNOV THEOREM

Lyapunov's theorem may be interpreted as a global result in the following way. The nonemptiness of L(A) is equivalent to the existence of a fixed positive definite matrix P such that

$$-\mathrm{Re}(x^*PAx) > 0$$

for all $0 \neq x \in C^n$. Thus one P must work for all (nonzero) x. It turns out that this global condition may be relaxed. Analogous to the "global" cone L(A), we define the "local" cone

$$L(A,x) = \{P^* = P > 0 : -\mathrm{Re}(x^*PAx) > 0\} \qquad (7)$$

We may then prove a local analog of Lyapunov's theorem (compare to Remark 1).

THEOREM 3. The n-by-n matrix A is stable if and only if $L(A,x) \neq \emptyset$ for all $0 \neq x \in C^n$.

Proof. This theorem is implied by the observation that $L(A,x) = \emptyset$ if and only if $Ax = \lambda x$ for some λ with $Re(\lambda) \geq 0$. First suppose that $Ax = \lambda x$ and $Re(\lambda) \geq 0$. Then, $Re(x^*PAx) = Re(x^*P(\lambda x)) = Re(\lambda)\cdot(x^*Px) \geq 0$, for all positive definite P. Thus $L(A,x) = \emptyset$. On the other hand, suppose $x \in C^n$ is such that $Ax \neq \lambda x$ for any complex number λ. Then it suffices to note that for x and y (=Ax), which are linearly independent, a $P^* = P > 0$ may always be constructed so that $Re(x^*Py) < 0$. To see this, pick S, non-singular, so that

$$Sy = \begin{bmatrix} 1 \\ 0 \\ \cdot \\ \cdot \\ \cdot \\ 0 \end{bmatrix} \quad \text{and } Sx = \begin{bmatrix} x_1 \\ x_2 \\ 0 \\ \cdot \\ \cdot \\ \cdot \\ 0 \end{bmatrix}$$

where $x_2 \neq 0$ and $Re(x_1) < 0$. Then $S^*S \in L(A,x)$, which is, therefore, nonempty. This completes the proof.

Remark 3. Since $L(A) = \cap\{L(A,x) : 0 \neq x \in C^n\}$, it follows from Remark 1 and Theorem 3 that $L(A,x) \neq \emptyset$ for all $0 \neq x \in C^n$ if and only if $L(A) \neq \emptyset$.

THE STABILITY OF $A + \alpha B$

A rather stronger condition than that the sum be stable may now be characterized for the pair A,B of stable matrices. The following two lemmas may be proved in a straightforward manner.

LEMMA 1. For all $\alpha > 0$, all $A \in M_n(C)$, and all $x \in C^n$, $L(\alpha A,x) = L(A,x)$.

LEMMA 2. For all $A,B \in M_n(C)$ and all $x \in C^n$, $L(A,x) \cap L(B,x) \subseteq L(A+B,x) \subseteq L(A,x) \cup L(B,x)$.

The principal result is that the analog of Theorem 2 now becomes a characterization.

THEOREM 4. Suppose A and B are stable. Then $A + \alpha B$ is stable for all $\alpha > 0$ if and only if $L(A,x) \cap L(B,x) \neq \emptyset$ for all $0 \neq x \in C^n$.

Proof. Suppose $L(A,x) \cap L(B,x) \neq \emptyset$ for all $0 \neq x \in C^n$. Given $\alpha > 0$, for any $0 \neq x \in C^n$ we have, by Lemmas 1 and 2, that $L(A+\alpha B,x) \neq \emptyset$. Thus, by Theorem 1, $A + \alpha B$ is stable.

Conversely, suppose that $A + \alpha B$ is stable for all $\alpha > 0$, that $0 \neq x \in C^n$ is arbitrary and that $P \in L(A,x)$ and $Q \in L(B,x)$. Then there is some $\alpha_1 > 0$ such that $P \in L(A+\alpha B,x)$ for $0 < \alpha < \alpha_1$, and there is some $\alpha_2 > 0$ such that $Q \in L(A+\alpha B,x)$ for $\alpha > \alpha_2$. Now $L(A+\alpha B,x)$ is a continuous, set valued function of $\alpha > 0$ and is nonempty for all $\alpha > 0$ because of Theorem 1. Also, $L(A+\alpha B,x) \subseteq L(A,x) \cup L(B,x)$ for all $\alpha > 0$, and $L(A,x)$ and $L(B,x)$ are open sets. If it were true that $L(A,x) \cap L(B,x) = \emptyset$, then either (i) $L(A+\alpha B,x) \subseteq L(A,x)$ for all $\alpha > 0$, or, (ii) $L(A+\alpha B,x) \subseteq L(B,x)$ for all $\alpha > 0$. However, these cases are both impossible, since $L(A,x) \cap L(A+\alpha B,x) \neq \emptyset$ for $0 < \alpha < \alpha_2$ and $L(B,x) \cap L(A+\alpha B,x) \neq \emptyset$ for $\alpha > \alpha_2$. We conclude that $L(A,x) \cap L(B,x) \neq \emptyset$, which completes the proof.

Remark 4. In the context of Theorem 4 it should be noted that none of the following three similarly attractive statements about stable A,B is valid, since 2 by 2 counterexamples may be constructed in each case:

1. $A + B$ is stable if and only if $L(A) \cap L(B) \neq \emptyset$
2. $A + \alpha B$ is stable for all $\alpha > 0$ if and only if $L(A) \cap L(B) \neq \emptyset$
3. $A + B$ is stable if and only if $L(A,x) \cap L(B,x) \neq \emptyset$ for all $0 \neq x \in C^n$.

REFERENCES

1. K. J. Arrow, Stability Independent of Adjustment Speed, in Trade, Stability, and Macroeconomics, Academic, New York, 1973.
2. K. J. Arrow and F. M. Hahn, General Competitive Analysis, Holden Day, San Francisco, 1971.
3. K. J. Arrow, H. D. Block and L. Hurwicz, On the Stability of the Competitive Equilibrium II, Econometrica 27 (1959), 89-109.

4. C. S. Ballantine, Stabilization by a Diagonal Matrix, Proc. Amer. Math. Soc. 25 (1970), 728-734.

5. R. Bellman, Introduction to Matrix Analysis, McGraw-Hill, New York, 1960.

6. C. R. Johnson, Sufficient Conditions for D-Stability, J. Econ. Theory 9 (1974), 53-62.

7. C. R. Johnson, A Characterization of the Nonlinearity of D-Stability, J. Math. Econ. 2 (1975), 87-91.

8. C. R. Johnson, Stable Matrices with Diagonal Lyapunov Solutions, to appear.

9. C. R. Johnson, A Lyapunov Theorem for Angular Cones, J. Research NBS 78B (1974), 7-10.

10. C. R. Johnson, A Local Lyapunov Theorem and the Stability of Sums, Lin. Alg. and Appl., to appear.

11. R. E. Kuenne, The Theory of General Economic Equilibrium, Princeton University, Princeton, 1963.

12. A. M. Lyapunov, Problème Général de la Stabilité du Mouvement, Ann. Math. Studies, No. 17, Princeton University, Princeton, 1947.

13. E. N. Onwuchekwa, Stability of Differential Equations with Applications to Economics, Thesis, Brown University 1975.

14. A. Takayama, Mathematical Economics, Dryden, Hinsdale, 1974.

15. O. Taussky Todd, A Recurring Theorem on Determinants, Am. Math. Monthly 56 (1949), 672-676.

Chapter 11

A NONCONTINUATION CRITERION FOR AN N-TH ORDER EQUATION WITH A RETARDED ARGUMENT

W. E. MAHFOUD*

Department of Mathematics
Southern Illinois University
Carbondale, Illinois

INTRODUCTION

In [1] Burton and Grimmer discussed noncontinuation of solutions of the second order ordinary differential equation

$$x''(t) + a(t)f(x(t)) = 0 \tag{A}$$

when $a(t)$ becomes negative at a point and $xf(x) > 0$ for $x \neq 0$, and gave the following result.

THEOREM A. Suppose $a(t_1) < 0$ for some $t_1 > 0$. If either

(i) $\int_0^{\infty} [1 + F(x)]^{-1/2} dx < \infty$, or

(ii) $\int_0^{-\infty} [1 + F(x)]^{-1/2} dx > -\infty$, where $F(x) = \int_0^x f(s)ds$,

then (A) has solutions which are not continuable to ∞.

*Present address: Department of Mathematics, Murray State University, Murray, Kentucky.

They also proved the converse of Theorem A and extended it later in [2] to the delay differential equation

$$x''(t) + a(t)f(x(q(t))) = 0 \tag{B}$$

when $a(t)$ becomes negative at a point, f is nondecreasing, and $xf(x) > 0$ for $x \neq 0$.

It is unknown whether or not Theorem A is extendable to Equation (B).

Recently, Burton [3] extended Theorem A to the n-th order ordinary differential equation

$$x^{(n)}(t) + a(t)f(x(t)) = 0$$

where $n \geq 2$.

MAIN RESULT

In this chapter we extend Theorem A to the differential equation

$$x^{(n)}(t) + a(t)f(x(t),x(q(t))) = 0 \tag{1}$$

where $n \geq 2$, $a : [0,\infty) \to R$, $R = (-\infty,+\infty)$, $q : [0,\infty) \to R$, and $f : R \times R \to R$.

We assume $a(t)$, $q(t)$, and $f(x,y)$ are continuous, $q(t) \leq t$ for all $t \geq 0$, and $f(x,y) > 0$ when $x > 0$ and $y > 0$ while $f(x,y) < 0$ when $x < 0$ and $y < 0$.

Following El'sgol'ts [4], for any $t_0 \geq 0$, we let $E_{t_0} = \{s : s = q(t) \leq t_0 \text{ for } t \geq t_0\} \cup \{t_0\}$. By a solution of (1) at t_0 is meant a function $x : E_{t_0} \cup [t_0,t_1] \to R$, for some $t_1 > t_0$, which satisfies (1) for all $t \in [t_0,t_1]$. Given a continuous function $\phi : E_{t_0} \to R$ and constants $c_1, \ldots, c_{n-1}$, there exists a solution $x(t)$ of (1) at t_0 with the property that $x(t) = \phi(t)$ for all $t \in E_{t_0}$ and $x^{(i)}(t_0) = c_i$ for $i = 1, \ldots, n-1$. A solution $x(t)$ of (1) at t_0 is said to be continuable if $x(t)$ exists for all $t \geq t_0$; otherwise, it is said to be noncontinuable.

We define $F_1(x) = \int_0^x f(s,s)ds$ and $F_{n+1}(x) = \int_0^x F_n(s)ds$ for $n = 1,2,3,\ldots$. It is clear from the definition of $F_n(x)$ that $(-1)^n F_{n-1}(x) > 0$ for $x < 0$ and for $n = 2,3,\ldots$. Let

$Q_1 = \{(x,y) : x \geq 0 \text{ and } y \geq 0\}$ and $Q_3 = \{(x,y) : x \leq 0 \text{ and } y \leq 0$. Define a function $f^* : Q_1 \cup Q_3 \to R$ such that for $x \leq 0$ and $y \leq 0$ we have $f^*(x,y) = f(x,y)$ and $f^*(-x,-y) = -f^*(x,y)$.

Consider the equation

$$x^{(n)}(t) + a(t)f^*(x(t),x(q(t))) = 0 \tag{2}$$

It is clear, from the definition of f^*, that f^* is continuous in $Q_1 \cup Q_3$, and if x and y have the same sign, then $f^*(x,y)$ has that sign. Also, if $x(t)$ is a solution of (2) such that $x(t)x(q(t)) > 0$, then $-x(t)$ is also a solution of (2). Moreover, $x(t) < 0$ with $x(t)x(q(t)) > 0$ is a solution of (2) if and only if $x(t)$ is a solution of (1).

We define $F^*_1(x) = \int_0^x f^*(s,s)ds$ and $F^*_{n+1}(x) = \int_0^x F^*_n(s)ds$ for $n = 1,2,3,\ldots$. Then $F^*_{n-1}(x)$ is even when n is even and odd when n is odd, and hence

$$(-1)^n F_{n-1}(x) = (-1)^n F^*_{n-1}(x) = F^*_{n-1}(-x) \tag{3}$$

for all $x < 0$.

NOTATION. For $d > 0$, we write

$R_d^+ = \{(x,y) : x \geq y \geq d\}$,

$R_d^- = \{(x,y) : x \leq y \leq -d\}$,

$C = \{f : R \times R \to R : f \text{ is continuous}\}$, and,

for any set $T \subset R \times R$,

$C_D(T) = \{f \in C : f$ is nonincreasing with respect to y for every fixed x whenever $(x,y) \in T\}$

THEOREM. Suppose $a(t_1) < 0$ for some $t_1 \geq 0$ and there is $d > 0$ such that either

(i) $\int_0^\infty [1 + F_{n-1}(x)]^{-1/n} dx < \infty$ and $f \in C_D(R_d^+)$, or

(ii) $\int_0^{-\infty} [1 + (-1)^n F_{n-1}(x)]^{-1/n} dx > -\infty$ and $f \in C_D(R_d^-)$.

Then (1) has noncontinuable solutions.

Proof. Suppose $\int_0^\infty [1 + F_{n-1}(x)]^{-1/n} dx < \infty$ and $f \in C_D(R_d^+)$; then for any given $\varepsilon > 0$ there exists $x_1 \geq d$ such that

$$\int_{x_1}^\infty [1 + F_{n-1}(x)]^{-1/n} dx < \varepsilon.$$

Since $a(t)$ is continuous and $a(t_1) < 0$, there exists $t_2 > t_1$ and positive constants m and M such that $-M \leq a(t) \leq -m$ for all $t \in [t_1, t_2]$.

Let $C_1, C_2, \ldots, C_{n-1}$ be positive constants to be determined. Let $x(t)$ be a solution of (1) such that $x(t) = x_1$ on E_{t_1} and $x^{(i)}(t_1) = C_i$ for $i = 1,2,\ldots,n-1$. We propose to show that, for some choice of the C_i's, $x(t)$ does not exist on $[t_1, t_2]$. If $x(t)$ exists on $[t_1, t_2]$, then it is clear that $x^{(i)}(t)$, $i = 0,1,\ldots,n-1$, are increasing on $[t_1, t_2]$. Since $q(t) \leq t$ for all $t \geq 0$, then $d \leq x_1 \leq x(q(t)) \leq x(t)$ for all $t \in [t_1, t_2]$. Since $f \in C_D(R_d^+)$, then $f(x(t), x(q(t))) \geq f(x(t), x(t))$ for all $t \in [t_1, t_2]$ and hence, by (1), we have $x^{(n)}(t) \geq -a(t)f(x(t),x(t)) \geq mf(x(t),x(t))$ for all $t \in [t_1, t_2]$. Multiply both sides of this inequality by $x'(t)$ and integrate from t_1 to $t \in [t_1, t_2]$ to get

$$\int_{t_1}^t x^{(n)}(s)x'(s)ds \geq m\int_{t_1}^t f(x(s),x(s))x'(s)ds$$

$$= m\int_{x_1}^{x(t)} f(s,s)ds$$

$$= m[F_1(x(t)) - F_1(x_1)]$$

An integration by parts on the left-hand side yields

$$x^{(n-1)}(t)x'(t) - x^{(n-1)}(t_1)x'(t_1) - \int_{t_1}^t x^{(n-1)}(s)x''(s)ds$$

$$\geq mF_1(x(t)) - mF(x_1)$$

As $x^{(i)}(t_1) = C_i$ and $x^{(i)}(t) > 0$ for $i = 1,\ldots,n-1$ and $t \in [t_1, t_2]$, we obtain

$$x^{(n-1)}(t)x'(t) \geq C_{n-1}C_1 - mF_1(x_1) + mF_1(x(t))$$

Choose $C_{n-1} = mF_1(x_1)/C_1$ so that $x^{(n-1)}(t)x'(t) \geq mF_1(x(t))$ for all $t \in [t_1, t_2]$. Multiply both sides of this inequality by $x'(t)$ and proceed as above to get

$$\int_{t_1}^{t} x^{(n-1)}(s)[x'(s)]^2 ds \geq m\int_{t_1}^{t} F_1(x(s))x'(s)ds$$

$$= m[F_2(x(t)) - F_2(x_1)]$$

and hence $x^{(n-2)}(t)[x'(t)]^2 \geq C_{n-2}C_1^{\,2} - mF_2(x_1) + mF_2(x)]$. Choose $C_{n-2} = mF_2(x_1)/C_1^{\,2}$ so that $x^{(n-2)}(t)[x'(t)]^2 \geq mF_2(x(t))$ for all $t \in [t_1, t_2]$. Continue this process (n-1) times and choose $C_i = mF_{n-i}(x_1)/C_1^{\,n-i}$, $i = 2,\ldots,n-1$, to get $x''(t)[x'(t)]^{n-2} \geq mF_{n-2}(x(t))$ for all $t \in [t_1, t_2]$. Multiply both sides by $x'(t)$ and integrate from t_1 to t to obtain $[x'(t)]^n - C_1^{\,n} \geq nmF_{n-1}(x(t)) - nmF_{n-1}(x_1)$ and hence $x'(t) \geq [k + mnF_{n-1}(x(t))]^{1/n}$, where $k = C_1^{\,n} - nmF_{n-1}(x_1)$. Choose $C_1 \geq [mn(1 + F_{n-1}(x_1))]^{1/n}$; then $k \geq mn$ and hence $x'(t) \geq k_1[1 + F_{n-1}(x(t))]^{1/n}$ where $k_1 = (mn)^{1/n}$. Thus $k_1 dt \leq [1 + F_{n-1}(x(t))]^{-1/n}dx(t)$. Integrate from t_1 to t to get

$$k_1(t - t_1) \leq \int_{x_1}^{x(t)} [1 + F_{n-1}(s)]^{-1/n}ds < \varepsilon \text{ for all } t \in [t_1, t_2].$$

Choose $\varepsilon = k_1(t_2 - t_1)$; then we obtain $t - t_1 < t_2 - t_1$ for all $t \in [t_1, t_2]$. This is a contradiction. Thus $x(t) \to \infty$ before t reaches t_2.

Now, suppose $\int_0^{-\infty} [1 + (-1)^n F_{n-1}(x)]^{-1/n}dx > -\infty$ and $f \in C_D(R_d^{\,-})$; then, by (3), we obtain $\int_0^{\infty}[1 + F^*_{n-1}(x)]^{-1/n}dx < \infty$. As $f^* \in C_D(R_d^{\,+})$, then by the above proof, (2) has a noncontinuable solution $y(t) > 0$. Let $x(t) = -y(t)$, then $x(t)$ is a solution of (1) which is not continuable. The proof is now complete.

EXAMPLE. The equation

$$x''(t) + a(t)\{x^3(t) + x(t/2)[1 + x^2(t/2)]^{-1}\} = 0$$

where $a(t) = -2[4 + (t-2)^2][4 + (t-2)^2 + 2(t-2)(t-1)^3]^{-1}$ if $0 \le t \le 1$ and $a(t) = -2$ if $t \ge 1$, has $x(t) = (t-1)^{-1}$ as a solution on $[0,1)$.

REFERENCES

1. T. A. Burton and R. Grimmer, On continuability of solutions of second order differential equations, Proc. Amer. Math. Soc. 29 (1971), 277-283.
2. T. A. Burton and R. Grimmer, Oscillation, continuation, and uniqueness of solutions of retarded differential equations, Trans. Amer. Math. Soc. 179 (1973), 193-209.
3. T. A. Burton, Noncontinuation of solutions of differential equations of order n, unpublished.
4. L. E. El'sgol'ts, Introduction to the theory of differential equations with deviating arguments, Holden Day, San Francisco, 1966.

Chapter 12

NEGATIVE ESCAPE TIME FOR SEMIDYNAMICAL SYSTEMS

ROGER C. McCANN

Department of Mathematics
Case Western Reserve University
Cleveland, Ohio

The purpose of this chapter is to investigate the concept of negative escape time for semidynamical systems.

R^+ will denote the non-negative reals. A semidynamical system on a topological space X is a continuous mapping $\Pi : X \times R^+ \to X$ such that

$$\Pi(x,0) = x \text{ for all } x \in X$$

$$\Pi(\Pi(x,t),s) = \Pi(x,t+s) \text{ for all } x \in X \text{ and } s,t \in R^+$$

If $A \subset X$ and $B \subset R^+$, then $\Pi(A,B)$ and $F(A,B)$ will denote the sets $\{\Pi(x,t) : x \in A,\ t \in B$ and $\{y : \Pi(y,t) \in A \text{ for some } t \in B\}$ respectively. A point $x \in X$ is called a start point if $x \notin \Pi(y,(0,\infty))$ for any $y \in X$.

Let Π and ρ be semidynamical systems on topological spaces X and Y respectively. Π is said to be isomorphic to ρ in the sense of Gottschalk and Hedlund (abbreviated as GH-isomorphic) if there exists a homeomorphism $h : X \to Y$ and a continuous mapping $\phi : X \times R^+ \to R^+$ such that

$$\phi(x,0) = 0 \text{ for each } x \in X$$

$$\phi(X,\cdot) : R^+ \to R^+ \text{ is a homeomorphism for each } x \in X$$

$$h(\Pi(x,t)) = \rho(h(x),\phi(x,t)) \text{ for each } (x,t) \in X \times R^+$$

Isomorphisms of local dynamical systems have been studied in depth by T. Ura [6,7]. We will say that Π can be embedded into ρ if there exists a homeomorphism h of X onto a subset of Y such that $h(\Pi(x,t)) = \rho(h(x),t)$ for every $(x,t) \in X \times R^+$.

Henceforth, Π will denote a semidynamical system on a Hausdorff space X. Intuitively, the negative escape time of a point $x \in X$ should be the minimal time length of all negative trajectories through x. Negative trajectories are defined and discussed at length in [1].

Since only an intuitive concept of negative trajectory is required in this chapter, we omit a precise definition and refer the interested reader to [1]. In order to make the concept of negative escape time precise, we need to consider the set M of all negative trajectories through x which originate at start points, and the set N_x of negative trajectories through x which do not originate at start points. We first define a negative escape time with respect to each set. Set $m(x) = \inf \{t \geq 0 : \Pi(y,t) = x$ for some start point $y\}$ if $M_x \neq \emptyset$ and $m(x) = +\infty$ if $M_x = \emptyset$. Set $n(x) = \inf \{t \geq 0 :$ there exist sequences $\{t_i\}$ in R^+ and $\{x_i\}$ in X such that $t_i \to t^-$, $\Pi(x_i,t_i) = x$, $x_i \in \Pi(x_i,R^+)$, and $\{x_i\}$ has no convergent subsequence$\}$ if $N_x \neq \emptyset$ and $n(x) = +\infty$ if $N_x = \emptyset$.

DEFINITION. The negative escape time, $N(x)$, of $x \in X$ is given by $N(x) = \min \{n(x), m(x)\}$. If $M \subset X$, $N(M) = \inf \{N(x) : x \in M\}$.

It should be noted that the negative escape time of $x \in X$ is with respect to all negative trajectories through x and not just a particular trajectory through x.

THEOREM 1. Let Π be a semidynamical system on a Hausdorff space X. Then Π can be embedded into a semidynamical system ρ on a Hausdorff space Y such that $N(y) = +\infty$ for all $y \in Y$. Moreover, ρ is minimal in the sense that if Π can also be embedded into a semidynamical system ρ' on a space Y', then ρ can be embedded into ρ'. Also, if $h : X \to Y$ is an embedding of Π into ρ, then $h(X)$ is positively invariant and if $y \in Y$, then $\Pi(y,t) \in h(X)$ for some $t \in R^+$.

In the proof of Theorem 1 (see McCann [4]), the space Y is constructed. Unfortunately, topological properties of X do not seem to be inherited by Y. In fact, the construction of Y is so complicated that it is difficult to determine any topological properties of Y. However, in certain circumstances it is possible to choose Y as X.

THEOREM 2. If there exists a continuous function $f : X \to (0,1]$ such that $f(x) \leq N(x)$ for all $x \in X$, then Π is GH-isomorphic to a semidynamical system ρ on X which has infinite negative escape time for each $x \in X$.

The condition $f(x) \leq N(x)$ imposes some type of continuity property on $N(\cdot)$. The following examples show that if Π has start points ($M_X \neq \emptyset$) or if X is not locally compact, then $N(\cdot)$ does not possess any "nice" continuity properties. In each example, Π will be the semidynamical system indicated in the diagram where $\Pi(x,t)$ is the point a distance t from x along the trajectory through x.

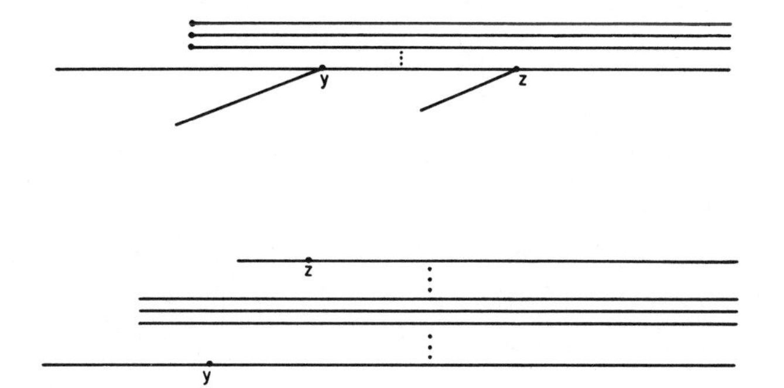

In each example there are sequences $\{y_i\}$, $y_i \to y$, and $\{z_i\}$, $z_i \to z$, such that $\limsup_{i \to \infty} N(y_i) < N(y)$ and $\liminf_{i \to \infty} N(z_i) > N(z)$. In light of these examples, we will restrict our attention to the situation that the semidynamical system Π has no start points and the phase space X is locally compact.

THEOREM 3. Let X be locally compact and Π have no start points. Then the negative escape time, $N(\cdot)$, is a lower semicontinuous function, i.e., $\liminf_{y \to x} N(y) \geq N(x)$ for all $x \in X$.

Since $N(\cdot)$ is a lower semicontinuous function, so is the function $g : X \to R^+$ defined by $g(x) = \min \{1, N(x)\}$. Since there are no start points, $N(x) > 0$ and, hence, $g(x) > 0$ for all $x \in X$. The constant function $h(x) = 0$ for all $x \in X$ is an upper semicontinuous function such that $h(x) < g(x)$. If X is locally compact and Lindelöf, then, by a well-known result of Dowker [2], there is a continuous $f : X \to R^+$ such that $0 = h(x) < f(x) < g(x) \leq N(x)$ for all $x \in X$. Theorem 2 yields

THEOREM 4. Let Π have no start points and X be locally compact and Lindelöf. Then Π is GH-isomorphic to a semidynamical system ρ on X which has infinite negative escape time for each $x \in X$.

It is well-known that a dynamical system on a locally compact space X can be extended to a dynamical system on the one point compactification X^* of X. This is not always possible for a semidynamical system (see [3; Chapter VI, Section 3.15]). Let $X^* = X \cup \{\infty\}$ be the one point compactification of the locally compact space X and define $\Pi^* : X^* \times R^+ \to X^*$ by

$$\Pi^*(x,t) = \begin{cases} \Pi(x,t) & \text{if } x \in X \\ \infty & \text{if } x = \infty \end{cases}$$

THEOREM 5. Let X be locally compact and Π have no start points. Then Π can be extended to the semidynamical system Π^* on X^* if and only if $N(x) = +\infty$ for every $x \in X$.

Since a semidynamical system on a manifold has no start points, [1; Theorem 11.8], the hypotheses of Theorems 3, 4, and 5 are satisfied whenever X is a manifold.

REFERENCES

1. N. Bhatia and O. Hajek, Local Semi-Dynamical Systems, Lect. Notes in Math. 90, Springer-Verlag, New York, 1969.

2. C. H. Dowker, On countably paracompact spaces, Canadian J. Math. 3 (1951), 219-244.

3. O. Hajek, Dynamical Systems in the Plane, Academic, New York, 1968.

4. R. McCann, An embedding theorem for semidynamical systems, Funkcial. Ekvac. 18 (1975), 23-34.

5. R. McCann, Negative escape time in semidynamical systems, Funkcial. Ekvac., to appear.

6. T. Ura, Isomorphism and local characterization of local dynamical systems, Funkcial. Ekvac. 12 (1969), 99-122.

7. T. Ura, Local isomorphism and local parallelizability in dynamical system theory, Math. Syst. Theory 3 (1969), 1-16.

Chapter 13

ON THE UNIFORM ASYMPTOTIC STABILITY OF THE LINEAR NONAUTONOMOUS EQUATION $\dot{x} = -P(t)x$ WITH SYMMETRIC POSITIVE SEMI-DEFINITE MATRIX P(t)

A. P. MORGAN* AND K. S. NARENDRA
Department of Engineering and Applied Science
Yale University
New Haven, Connecticut

INTRODUCTION

The ordinary differential equation $\dot{x} = -P(t)x$ where P(t) is symmetric positive semi-definite time-varying matrix arises often in mathematical control theory. (See, for example, Narendra and McBride [7; p. 34], Lion [5; p. 1837], and Sondhi and Mitra [10; p. 5].)

In this chapter we consider the stability properties (in the sense of Lyapunov) of the equilibrium state $x \equiv 0$. Since for $V(x) = x^T x$, $\dot{V}(x) \leq 0$, the origin is uniformly stable. However (uniform) asymptotic stability does not generally hold unless P(t) is positive definite. The semi-definite case arises much more frequently in practice than the definite one, and the main effort in this paper is directed towards finding conditions characterizing uniform asymptotic stability in such a case.

---

*Present address: Department of Mathematics, University of Miami, Coral Gables, Florida

For applications, uniform asymptotic stability is important because this property is preserved under perturbations (see Hale [2; Theorems 2.3, 2.4 and 5.2]). On the other hand, this "structural stability" is not necessarily possessed by (non-uniform) asymptotically stable systems (see Hale [2; p. 87]). Note also that since $\dot{x} = -P(t)x$ is linear, all stability properties are global.

The principal results are stated in Theorem 1 and Lemma 2. The following Theorem, which is a part of Theorem 1, gives a simple and complete characterization of uniform asymptotic stability and is illustrative of the type of result derived in this chapter.

THEOREM. Suppose $P(t)$ is a symmetric positive semi-definite matrix of bounded piecewise continuous functions. Then the equation

$$\dot{x} = -P(t)x \tag{1}$$

is uniformly asymptotically stable if and only if there are real numbers $a > 0$ and b such that

$$\int_{t_0}^{t} |P(s)\cdot w|\,ds \geq a(t - t_0) + b$$

for all $t \geq t_0 \geq 0$ and all fixed unit vectors w.

In the next section we discuss some examples. In the last section the principal results are stated, and a lemma useful in showing uniform asymptotic stability for other classes of linear and nonlinear systems of equations is given. For proofs of the results presented here, see Morgan and Narendra [6].

PRELIMINARY DISCUSSION

Before stating all our main results, we will discuss some implications of the theorem above. Our discussion divides naturally into five parts (a,b,c,d, and e below). First, however, we state the following.

DEFINITION. The equilibrium state $x \equiv 0$ of the uniformly stable differential equation $\dot{x} = f(x,t)$ is uniformly asymptotically stable (u.a.s.) if for some $\varepsilon_1 > 0$ and all $\varepsilon_2 > 0$ there is a $T = T(\varepsilon_1, \varepsilon_2) > 0$ such that if $x(t)$ is a solution and $|x(t_0)| < \varepsilon_1$, then $|x(t)| < \varepsilon_2$ if $t \geq t_0 + T$. If T depends on t_0, then $\dot{x} = f(x,t)$ is (non-uniformly) asymptotically stable (a.s.).

(a) If $P(t) = P$ is a constant nxn matrix, then the following are equivalent. (i) Equation (1) is u.a.s. (ii) Equation (1) is a.s. (iii) P has rank n.

(b) Let $\lambda(t)$ denote the eigenvalue of minimal length of $P(t)$. Then u.a.s. holds if there are $a > 0$ and b such that

$$\int_{t_0}^{t} |\lambda(s)|ds \geq a(t - t_0) + b$$

for all $t \geq t_0$. In particular, if $P(t)$ has (maximal) rank n for all t and $\lambda(t)$ is bounded above zero or periodic, then $\dot{x} = -P(t)x$ is u.a.s. Thus if $P(t)$ is rank n and periodic, then u.a.s. holds. However,

$$\int_{t_0}^{t} |\lambda(s)|ds \geq a(t - t_0) + b$$

is not necessary but only sufficient. This will be clear from the discussion of the 2x2 rank 1 case in part (c) below.

(c) Suppose there is $u : [0,\infty) \to R^2$ such that

$$P(t) = u(t) \cdot u(t)^T = \begin{bmatrix} u_1^2 & u_1u_2 \\ u_1u_2 & u_2^2 \end{bmatrix}$$

The eigenvalues of $P(t)$ are then $|u(t)|^2 = u_1(t)^2 + u_2(t)^2$ and 0. Now

$$\dot{x} = -P(t) \cdot x$$

becomes $\dot{x} = -\langle u(t), x\rangle \cdot u(t)$ where $\langle , \rangle$ denotes the canonical inner product on R^2. Thus the condition

$$\int_{t_0}^{t} |P(s)w|ds = \int_{t_0}^{t} |\langle u(s), w\rangle||u(s)|ds \geq a(t - t_0) + b$$

for fixed unit vectors w requires that both $|\langle u(s),w\rangle|$ and $|u(s)|$ "not get too small for too long." Thus, u(s) must change direction uniformly so that its inner product with any fixed direction w does not converge too quickly to zero, and u(s) itself must not converge too quickly to zero. To further illustrate this, consider the following explicit examples.

(d) Let $e_1 = (1,0)$ and $e_2 = (0,1)$. Define vectors u(t) and u'(t) to alternate between e_1 and e_2 according to the following formulas.

$$u(t) = \begin{cases} e_1, & t \in [2n,2n+1) \\ e_2, & t \in [2n+1,2n+2) \end{cases} \qquad n = 0,1,2,\ldots$$

$$u'(t) = \begin{cases} e_1, & t \in [0,1) \cup [2,4) \cup [5,8) \cup \ldots \\ e_2, & t \in [1,2) \cup [4,5) \cup [8,9) \cup \ldots \end{cases}$$

Now $\dot{x} = -u(t)u(t)^T x$ is u.a.s., because for at least half the time $|\langle u(s),w\rangle||u(s)| \geq \max\{|\langle e_1,w\rangle|,|\langle e_2,w\rangle|\} > 0$. But $\dot{x} = -u'(t)u'(t)^T x$ is not u.a.s. because u' spends longer and longer time in the e_1 direction. Solutions with initial conditions on the y-axis must wait longer and longer before they can go to zero. It is clear that

$$\int_{t_0}^{t} |\langle u'(s),e_2\rangle||u'(s)|ds = \int_{t_0}^{t} |\langle u'(s),e_2\rangle|ds$$

equals zero for longer and longer intervals and can dominate no linear function with positive slope. However, the above integral does go to infinity as $t \to \infty$, and this implies that $\dot{x} = -u'(t)u'(t)^T x$ is asymptotically stable.

(e) Consider the following final example. Let $u(t) = (1,t^{-1/2})$. Then

$$u(t)u(t)^T = \begin{bmatrix} 1 & t^{-1/2} \\ t^{-1/2} & \frac{1}{t} \end{bmatrix}$$

and $|\langle u(s),w\rangle||u(s)| = |w_1 + t^{-1/2}w_2||(1 + 1/t)^{1/2}|$. Thus for $w = (0,1)$, we require that

$$\int_{t_0}^{t} s^{-1/2}(1 + 1/s)^{1/2}ds \leq \int_{t_0}^{t} 2s^{-1/2}ds = 2s^{1/s}\Big|_{t_0}^{t}$$

dominate a linear function. But this is false. It is easy to confirm that if $u(t) = (1,t^{\alpha})$ where $\alpha < 0$, then $\dot{x} = -u(t)u(t)^T x$ is not u.a.s. It can be shown that such equations are not even a.s.

We close this section by noting that the comments made in (c), (d), and (e) clearly hold for the general nxn case.

PRINCIPAL RESULTS

If $P(t)$ is symmetric positive semi-definite, then there is a symmetric $u(t)$ such that $P(t) = u(t)^2 = u(t)u(t)^T$ (see Reed and Simon [9; p. 196]). We will usually assume $P(t)$ is in this form. As a special case we consider $P(t) = u(t)u(t)^T$ with $u(t)$ an nxk matrix with $k \leq n$. In this case, $u(t)u(t)^T$ can have at most rank k. In general, $u(t)$ is nxn but not necessarily of full rank. In fact, the rank of $u(t)u(t)^T$ may change with t. We do assume that $u(t)$ is piecewise continuous and uniformly bounded.

Letting $V(x) = x_1^2 + x_2^2 + \ldots + x_n^2$, we see that $\dot{V}(x) = -x^T P(t)x \leq 0$ for $\dot{x} = -P(t)x$. Thus the equation is easily seen to be uniformly stable. If $P(t)$ is constant or periodic, we have the well known result of LaSalle [3] by which, if V is not constant on any solution of $\dot{x} = -P(t)x$, asymptotic stability follows. This result breaks down for general non-autonomous $P(t)$. This can be seen as a result of the lack of an invariance property for the ω-limit set (see LaSalle [4]).

DEFINITION. Let ∂S_r denote a sphere of radius r about 0 and S_r a ball of radius r about 0. Thus $\partial S_r = \{x \in R^n | |x| = r\}$ and $S_r = \{x \in R^n | |x| \leq r\}$. By a conical neighborhood C^{α}_y for y we mean that α is an open subset of the unit sphere

$\partial S_1 \subseteq R^n$, $y/|y| \in \alpha$ or $-y/|y| \in \alpha$ if $y \neq 0$, and C^α_y is defined to be the union of all lines through 0 in R^n that intersect α. The width of C^α_y is defined to be the diameter of α. For simplicity, we sometimes omit the α and write C_y instead of C^α_y.

DEFINITION. By $f : [0,\infty) \to R^1$ piecewise continuous, we mean that there is a decomposition of $[0,\infty)$ into half-open intervals, $[0,\infty) = \bigcup_{n=1}^{\infty} [a_n, a_{n+1})$ such that the restriction $f|(a_n, a_{n+1})$ is continuous for all n.

The following theorem gives a characterization of uniform asymptotic stability for $\dot{x} = -P(t)x$. The statement of the theorem is followed by a key lemma and some remarks. Proofs will appear in Morgan and Narendra [6]. In reading the following material, the reader may find the case $u : [0,\infty) \to R^2$ an illuminating example.

THEOREM 1. Let $u : [0,\infty) \to R^n_k$ be a piecewise continuous and bounded function, where R^n_k denotes the space of real nxk matrices. (We identify R^n_1 and R^n.) Then the following are equivalent.

(i) $\dot{x} = -u(t)u(t)^T x$ is uniformly asymptotically stable.

(ii) There are real numbers $a > 0$ and b such that if $y \in R^n$ is a fixed unit vector, then

$$\int_{t_0}^{t} y^T u(s)u(s)^T y\,ds \geq a(t - t_0) + b$$

for all $t \geq t_0 \geq 0$.

(iii) There are real numbers $a > 0$ and b such that

$$\lambda_i \left[\int_{t_0}^{t} u(s)u(s)^T ds \right] \geq a(t - t_0) + b$$

for $i = 1,2,\ldots,n$ where λ_i denotes the ith eigenvalue of the nxn matrix

$$\int_{t_0}^{t} u(s)u(s)^T ds$$

(iv) Given y a unit vector in R^n, there is a conical neighborhood C_y for y and there are real numbers $a_y > 0$ and b_y such that

$$\int_{[t_0,t]-\Omega_y} |u(s)|^2 ds \geq a_y(t - t_0) + b_y$$

for all $t \geq t_0 \geq 0$ where $\Omega_y = \{t \in [0,\infty) | u(t)^{\perp} \cap C_y \neq 0\}$, and $u(t)^{\perp}$ = orthogonal complement of u(t) = kernel $(u(t)^T)$.

Part (iv) is more technical than the others and helps to bridge the gap between parts (i) and (ii) in the proof. It says, intuitively, that u(t) is bounded away from each unit direction for a sufficient part of time over any reasonably long period of time. However, it is formulated to say that $u(t)^{\perp}$ is bounded away from any unit direction, which is actually more to the point.

REMARK. We may replace the integral expression in (ii) by

$$\int_{t_0}^{t} |u(s)u(s)^T y| ds \geq a (t - t_0) + b$$

or by

$$\int_{t_0}^{t} |u(s)^T y| ds \geq a(t - t_0) + b$$

The theorem stated in the first section asserts the equivalence of parts (i) and (ii) of Theorem 1, except that only one of the three formulations of part (ii) (see the previous Remark) is given there. In practice, it would seem that the equivalence of parts (i) and (ii) would be the most useful implication of this theorem. We should also note that the equivalence of part (ii) and the eigenvalue condition, part (iii), is not hard to show.

REMARK. After the presentation of this chapter, it was pointed out to the authors that B. D. O. Anderson (Department of Electrical Engineering, Univ. of New Castle, Australia), for the case that P(t) is almost periodic, has established results from which it follows that (ii) implies (i).

The following key lemma will be applicable to many cases besides those discussed in this chapter. To indicate this, we present some corollaries after the statement of the lemma.

First we need a definition.

DEFINITION. A function $\phi : [0,\infty) \to [0,\infty)$ is said to belong to class K, $\phi \in K$, if it is continuous, strictly increasing, and $\phi(0) = 0$.

LEMMA 2. Let $f(x,t) : R^n \times [0,\infty) \to R^n$ be piecewise continuous with $f(0,t) = 0$ for all t. Assume:

(i) There is $\phi_1 \in K$ and an $\varepsilon > 0$ such that $|f(x,t) - f(y,t)| \leq \phi_1(|x - y|)$ for all x,y,t with $|x - y| < \varepsilon$.

(ii) There are real numbers $a > 0$ and b and $\phi_2 \in K$ such that

$$\int_{t_0}^{t} |f(x,s)|ds \geq \phi_2(|x|)[a(t - t_0) + b]$$

for all fixed $x \in R^n$ and $t \geq t_0 \geq 0$.

(iii) There is a continuously differentiable function $V : R^n \times [0,\infty) \to [0,\infty)$ and $\phi_3 \in K$ such that $\phi_3(|x|) \geq V(x,t) > 0$ if $x \neq 0$, $V(0,t) \equiv 0$, and $\dot{V}(x,t) \leq 0$ for all t,x where

$$\dot{V}(x,t) = \frac{\partial V}{\partial t}(x,t) + \nabla V(x,t) \cdot f(x,t)$$

(iv) There is a $\phi_4 \in K$ such that $-\dot{V}(x,t) \geq |f(x,t)|^2 \cdot \phi_4(|x|)$ for all $x \in R^n$, $t \in [0,\infty)$.

(v) The solution $x \equiv 0$ of the equation $\dot{x} = f(x,t)$ is uniformly stable.

Then the solution $x \equiv 0$ for the equation $\dot{x} = f(x,t)$ is uniformly asymptotically stable.

REMARKS. Condition (i) is satisfied if $f(x,t) = A(t)x$ and $|A(t)| \leq M$ for some constant M, all t. It is also satisfied if f is differentiable in x and its derivative with respect to x is bounded uniformly in t. Intuitively, something like condition (ii) seems necessary for u.a.s. However, it probably is not necessary as written. Since Lyapunov function converse theorems for uniform asymptotic stability exist, condition (iii) is very natural. (See Hale [2; Chapter X]). We know, from Krasovskii's theorem, that if $\dot{x} = A(t)x$ is u.a.s., then a quadratic Lyapunov function exists (see Narendra and Taylor [8; p. 62]). In this case, if $|A(t)|$ is uniformly bounded, it is easy to see that we can choose ϕ_3 to make condition (ii) hold. Thus, for $f(x,t)$ linear and V quadratic, condition (iv) is necessary for u.a.s. If there is a $\phi \in K$ such that $V(x,t) \geq \phi(|x|)$ for all x and t, then uniform stability (condition (v)) follows.

DEFINITION. $A \geq B$ means $A - B$ is positive semi-definite.

COROLLARY 3. If $f(x,t) = -P(t)x$, where $P(t)$ is a symmetric positive definite uniformly bounded matrix, and if there are real numbers $a > 0$ and b such that

$$\int_{t_0}^{t} |P(s)w|\,ds \leq a(t - t_0) + b$$

for all $t \geq t_0 \geq 0$ and all fixed unit vectors w, then $\dot{x} = -P(t)x$ is u.a.s.

Proof. Applying Lemma 2, conditions (i) and (ii) are immediate. Letting $V(x) = |x|^2$, we have $\dot{V}(x,t) = -x^TP(t)x \leq 0$ so conditions (iii) and (v) are also easy. Condition (v) follows because $0 \leq P(t) \leq I$ implies $P(t)^2 \leq P(t)$ for symmetric $P \geq 0$. (We may as well assume $P(t) \leq I$.) Thus $-\dot{V}(x,t) = x^TP(t)x \geq x^TP(t)^2x = |P(t)x|^2$.

COROLLARY 4. Suppose $\dot{x} = A(t)x$ is uniformly stable, $A(t)$ is uniformly bounded, and there are real numbers $a > 0$ and b such that

$$\int_{t_0}^{t} |A(s)w|\,ds \geq a(t - t_0) + b$$

for all $t \geq t_0 \geq 0$ and all unit vectors w. Assume there is a positive definite $Q(t)$ uniformly bounded such that

$$-(Q(t)A(t) + A(t)^TQ(t) + \dot{Q}(t)) \geq cA(t)^TA(t)$$

for all t where c is some positive constant. Then $\dot{x} = A(t)x$ is u.a.s.

Proof. Let $V(x,t) = x^T Q(t)x$. Then the result follows immediately from Lemma 2.

REFERENCES

1. L. Cesari, Asymptotic Behavior and Stability Problems in Ordinary Differential Equations, Third Ed., Springer-Verlag, New York, 1971.

2. J. K. Hale, Ordinary Differential Equations, Wiley, New York, 1969.

3. J. P. LaSalle, Asymptotic Stability Criteria, Proceedings of the Symposia in Applied Math., Hydrodynamic Instability, Vol. 13, Amer. Math. Soc., Providence, 1962, 299-307.

4. J. P. LaSalle, Stability theory for ordinary differential equations, Differential Equations 4 (1968), 57-65.

5. P. M. Lion, Rapid identification of linear and nonlinear systems, AIAA J. 5 (1967), 1835-1847.

6. A. P. Morgan and K. S. Narendra, On the uniform asymptotic stability of certain linear nonautonomous differential equations, SIAM J. Control, to appear.

7. K. S. Narendra and L. E. McBride, Multiparameter Self-optimizing Systems Using Correlation Techniques, IEEE Trans. Automatic Control, AC-9 (1964), 31-38.

8. K. S. Narendra and J. H. Taylor, Frequency Domain Criteria for Absolute Stability, Academic, New York, 1973.

9. M. Reed and B. Simon, Methods of Mathematical Physics 1: Functional Analysis, Academic, New York, 1972.

10. M. M. Sondhi and D. Mitra, New Results on the Performance of a Well-Known Class of Adaptive Filters, (Bell Laboratories, Murry Hill, NJ 07974), April 1975.

Chapter 14

SMOOTHING CONTINUOUS DYNAMICAL SYSTEMS ON MANIFOLDS

DEAN A. NEUMANN
Department of Mathematics
Bowling Green State University
Bowling Green, Ohio

INTRODUCTION

Let M be a C^∞ manifold and let $\phi : M \times R^1 \to M$ be a continuous flow (dynamical system) on M. We consider the question, posed by Hajek in [6], of the existence of a C^∞ flow ψ on M that is topologically equivalent to ϕ (we say that ϕ and ψ are topologically equivalent if there is a homeomorphism of M that takes orbits of ϕ onto orbits of ψ, preserving the natural orientation of the orbits). If there is such a C^∞ flow ψ we will say that ϕ is smoothable. The purpose of this chapter is to indicate several recent results on this smoothing problem. Complete proofs will appear in [14] and [15].

COMPLETELY UNSTABLE FLOWS

It is known that considerable restriction must be placed on (M,ϕ) in order to guarantee that ϕ is smoothable. For example, the non-ergodic C^1 flows on the 2-dimensional torus described by Denjoy [5] (also see Hartman [8; Chapter 7]), are known to be inequivalent to even C^2 flows. In this case the reason is the existence of an exotic non-wandering set. Since these examples can be embedded in flows on any manifold of higher dimension, some restriction must be placed on the non-wandering set $\Omega(\phi)$ of ϕ.

It is therefore natural to consider first flows (M,ϕ) in which $\Omega(\phi)$ is empty; such a flow is called completely unstable. Even in this case the dimension of M must be restricted to guarantee that ϕ is smoothable; we have in fact:

THEOREM 1. If M is a C^∞ m-manifold with $m \leq 3$ and ϕ is any completely unstable continuous flow on M, then ϕ is smoothable. For each $m \geq 4$, there is a completely unstable continuous flow ϕ_m on R^m that is not smoothable.

The first part of Theorem 1 is proved by reducing the smoothing problem for (M,ϕ) to the problem of imposing a C^∞-structure on the orbit space M/ϕ (here M/ϕ denotes the set of orbits of ϕ with the strongest topology in which the natural projection $\pi : M \to M/\phi$ is continuous). Under the assumptions of Theorem 1, M/ϕ is a non-separated (topological) (m-1)-manifold, i.e., a (not necessarily Hausdorff) space with a countable basis of open sets each homeomorphic to R^{m-1}. This may be seen as follows: if ϕ is a completely unstable continuous flow on an m-manifold M (of arbitrary dimension) then, through each point of M, there is a local section S of ϕ such that $S\cdot R^1 = \{\phi(s,t) : s \in S, t \in R^1\}$ is an open subset of M homeomorphic with $S \times R^1$ under (the inverse of) the restriction of ϕ (cf. [1; Theorem 2.12]). If $m \leq 3$ then we may assume that S is homeomorphic to R^{m-1} (cf. [6], [17] for $m = 2$; [4], [16] for $m = 3$). The first assertion of Theorem 1 then follows from the next two theorems and the fact that the C^∞ structure on a manifold of dimension $m \leq 3$ is unique up to diffeomorphism (Munkres [13]).

THEOREM 2. Let M be a (Hausdorff) topological m-manifold and let $\phi : M \times R^1 \to M$ be a completely unstable continuous flow on M. Then M can be given a C^∞ structure with respect to which ϕ is C^∞ if and only if M/ϕ can be given the structure of a C^∞ non-separated (m-1)-manifold.

THEOREM 3. Any non-separated m-manifold, with $m \leq 3$, admits a C^{∞} structure.

Note that there is no restriction on dimension in Theorem 2. The definition of C^{∞} structure for non-separated manifolds is just as in the case of Hausdorff manifolds. The dimension restriction in Theorem 3 is imposed by the proof. It is known that in each dimension $m \geq 5$, there are (Hausdorff) m-manifolds that do not admit any C^{∞} structure (this follows from results of Kirby and Sibenman [10]), but the remaining case ($m = 4$) is unknown and includes the difficult triangulation-smoothing problem for (Hausdorff) 4-manifolds.

The existence of non-smoothable completely unstable flows on R^m ($m \geq 4$) follows from the existence of non-Euclidean factors of R^m ($m \geq 4$). The first such example was given by Bing in [2]. In [3], Chewning uses Bing's example to construct a non-smoothable flow on R^4. Similar examples, which we describe briefly, exist in all higher dimensions: it is proved in [11] that, for each $m \geq 4$, there is a space X_m that is not locally Euclidean at any point, but with $X_m \times R^1 \simeq R^m$. The flow ϕ_m defined on R^m by

$$\phi_m((x,t),s) = (x,t + s) \qquad (x \in X_m;\ s,t \in R^1)$$

is completely unstable, but cannot be equivalent to even a C^1 flow. To see this observe that the orbit space R^m/ϕ of a continuous flow ϕ is an invariant of the topological equivalence class of ϕ. The orbit space of a completely unstable C^1 flow on R^m is known to be locally (m-1)-Euclidean (Markus [12; Theorem 3]), while $R^m/\phi_m \simeq X_m$ is not.

LOCALLY TRIVIAL, COMPLETELY UNSTABLE FLOWS

We will say that a continuous flow ϕ on an m-manifold M is locally trivial if, through each regular point in M, ϕ admits a local section that is homeomorphic to R^{m-1}. The examples of non-smoothable flows given in the preceding section depend on the fact that, in higher dimensions, a continuous flow need not be locally trivial. The next theorem shows that, for

4-manifolds, this is the only way in which a completely unstable flow can fail to be smoothable in some C^∞ structure. In higher dimensions the situation is more complicated.

THEOREM 4. Let M be a topological 4-manifold and let ϕ be a locally trivial, completely unstable, continuous flow on M. Then there is a C^∞ structure on M with respect to which ϕ is C^∞.

The conclusion does not necessarily imply that ϕ is smoothable, in the sense defined above, when M has a given C^∞ structure. We could make the stronger assertion if it were known that a C^∞ structure on a 4-manifold was unique up to diffeomorphism, but this question is still unresolved.

The analogous result fails for manifolds of dimension $m \geq 6$. We may construct counterexamples using the fact mentioned above that, for each $m \geq 5$, there is a topological m-manifold M_m that does not admit any C^∞ structure. The natural flow ϕ_m on $M_m \times R^1$, defined by $\phi_m(x,s,t) = (x,s+t)$, is then locally trivial and completely unstable. But, as $M_m \times R^1/\phi_m \simeq M_m$, Theorem 2 implies that ϕ_m cannot be C^∞ in any C^∞ structure on $M_m \times R^1$.

The remaining case ($m = 5$) appears to be very difficult.

SMOOTHING C^1 FLOWS

In contrast to the Denjoy examples (which are C^1, but inequivalent to C^2 flows) a completely unstable flow is smoothable if and only if it is equivalent to a C^1 flow.

THEOREM 5. Let M be a C^∞ manifold of arbitrary dimension, and let ϕ be a completely unstable C^1 flow on M. Then ϕ is smoothable.

This is the analogue of the well known fact that any maximal C^1 structure on a (Hausdorff) differentiable manifold contains a C^∞ structure. Theorem 5 follows from a slight strengthening of Theorem 2 and the following generalization of this fact.

THEOREM 6. Let M be a C^1 non-separated manifold. Then any maximal C^1 atlas on M contains a C^∞ atlas.

FLOWS WITH NON-WANDERING POINTS; FLOWS ON 2-MANIFOLDS

We next consider flows (M,ϕ) with $\Omega(\phi)$ nonempty. In this case the orbit space M/ϕ is almost never a manifold and the technique of the preceding sections is no longer applicable. However, in the simple case in which $\Omega(\phi)$ consists entirely of rest points, we can apply Theorem 1 as follows: let $\tilde{M}$ denote $M - \Omega(\phi)$ and let $\tilde{\phi}$ denote the restriction of ϕ to $\tilde{M}$. By Theorem 1, there is a C^∞ flow $\tilde{\psi}$ on $\tilde{M}$ that is topologically equivalent to $\tilde{\phi}$. It can be arranged that the homeomorphism of $\tilde{M}$ that realizes the equivalence is the restriction to $\tilde{M}$ of a homeomorphism of M that fixes each point of $\Omega(\phi)$. If $u : M \to [0,1]$ is a C^∞ function that vanishes exactly on $\Omega(\phi)$, and is C^∞ flat at $\Omega(\phi)$, then the fector field $u \cdot \frac{d}{dt} \psi : \tilde{M} \to T\tilde{M}$ extends to a C^∞ vector field $\chi : M \to TM$, whose induced flow is topologically equivalent to ϕ. We thus have the following extension of Theorem 1.

<u>THEOREM 1'.</u> If M is a C^∞ m-manifold with $m \leq 3$ and ϕ is any continuous flow on M such that $\Omega(\phi)$ consists entirely of rest points, then ϕ is smoothable.

Beyond this our results on flows with non-wandering points are restricted to compact 2-manifolds, but in this case are reasonably complete.

<u>THEOREM 7.</u> Suppose that ϕ is a continuous flow on the compact orientable 2-manifold M. Assume that ϕ has at most finitely many rest points and that any recurrent point of ϕ is periodic. Then ϕ is smoothable.

The Denjoy examples mentioned above show that the restriction on recurrent orbits is necessary.

Since there can be no non-periodic recurrence in the plane we have the following corollary.

<u>COROLLARY.</u> Any continuous flow on R^2 with at most finitely many rest points is smoothable.

The corollary was proved by Kaplan in [9] in the case of flows with no rest points.

The proof of Theorem 8 depends on a partial classification of flows with no non-periodic recurrence on 2-manifolds. It can be shown that, if (M,ϕ) satisfies the hypothesis of the theorem, and $\tilde{M}$ denotes the complement in M of the rest points of ϕ, then $\tilde{M}$ can be decomposed into a locally finite collection of closed, ϕ-invariant submanifolds, on each of which the restriction of ϕ is of a simple type (roughly, either a flow that admits a complete cross-section, or a flow that is completely unstable on the interior of the submanifold). We can construct C^∞ models of arbitrary flows of these simple types, in the latter case, using the results in the second section of this Chapter. We then construct a C^∞ model of $(\tilde{M}, \tilde{\phi} = \phi|_{\tilde{M}})$ by glueing up models of the submanifolds of the decomposition. The equivalence of $(\tilde{M}, \tilde{\phi})$ with this model induces a C^∞ structure on $\tilde{M}$ with respect to which $\tilde{\phi}$ itself is C^∞. Munkres' theorem on the uniqueness of a differentiable structure on a 2-manifold then implies that there is a flow ψ on $\tilde{M}$ that is C^∞ with respect to the given structure on $\tilde{M}$ and is topologically equivalent to $\tilde{\phi}$; ψ is easily extended to a C^∞ flow on M that is topologically equivalent to ϕ.

REFERENCES

1. N. P. Bhatia and G. P. Szegö, Stability Theory of Dynamical Systems, Springer-Verlag, New York, 1970.
2. R. H. Bing, The cartesian product of a certain non-manifold and a line is E^4, Ann. Math. 70 (1959), 399-412.
3. W. C. Chewning, A dynamical system on E^4 neither isomorphic nor equivalent to a differential system, Bull. Amer. Math. Soc. 80 (1974), 150-154.
4. W. C. Chewning and R. S. Owen, Local sections of flows on manifolds, Proc. Amer. Math. Soc. 49 (1975), 71-77.
5. A. Denjoy, Sur les courbes definies par les equations differentielles a la surface du tore, Journal de Mathematique (9) 11 (1932), 333-375.
6. O. Hajek, Dynamical Systems in the Plane, Academic, New York, 1968.
7. O. Hajek, Sections of dynamical systems in E^2, Czechoslovak Math. J. 15 (1965), 205-211.
8. P. Hartman, Ordinary Differential Equations, Wiley, New York, 1964.
9. W. Kaplan, Differentiability of regular curve families on the sphere, in Lectures in Topology, R. L. Wilder and W. L. Ayres, eds., Univ. of Mich., Ann Arbor, 1941.

10. R. Kirby and L. C. Siebenman, Some theorems on topological manifolds, in Manifolds--Amsterdam, 1970, Lect. Notes in Math., Vol. 197, N. H. Kuiper ed., Springer-Verlag, New York, 1971.

11. K. W. Kwun, Factors of M-space, Mich. Math. J. 9 (1962), 207-211.

12. L. Markus, Parallel dynamical systems, Topology 8 (1969), 47-57.

13. J. Munkres, Obstructions to smoothing piecewise differentiable homeomorphisms, Ann. Math. 72 (1960), 521-554.

14. D. Neumann, Smoothing continuous flows, J. Differential Equations, to appear.

15. D. Neumann, Smoothing continuous flows on 2-manifolds, to appear.

16. H. Whitney, Cross sections of curves in 3-space, Duke Math. J. 4 (1938), 222-226.

17. H. Whitney, Regular families of curves, Ann. Math. 34 (1933), 244-270.

Chapter 15

GLOBAL ASYMPTOTIC STABILITY OF SOME AUTONOMOUS THIRD ORDER SYSTEMS

R. REISSIG
Institut für Mathematik
Ruhr-Universität Bochum
Federal Republic of Germany

The intention of the present chapter is to point out that the invariance principle of J. P. LaSalle [6] which played an important role in the principal lectures of this conference, is not only interesting in connection with the theory of dynamical systems, but is particularly useful in solving practical problems. Applying the invariance principle we are able to derive stability criteria which only depend upon the nature of the considered dynamical system, and which are optimal, in a certain sense, whereas other procedures lead to stability conditions which are more or less artificial.

Let us refer to a recent paper of S. Kasprzyk [4] which is devoted to four nonlinear third order systems, and the purpose of which is the application of the well-known Hartman-Olech theorem on global asymptotic stability [3]. This procedure requires a subtle transformation of the considered differential system which, of course, influences the stability conditions. However, such a transformation becomes superfluous and the stability theorem of Kasprzyk can be improved considerably when the study of the systems is based on LaSalle's invariance principle.

Consider the following differential systems where the coefficients of the linear terms are assumed to be positive real numbers:

$$x''' + ax'' + bx' + f(x) = 0 \tag{1}$$

$$x''' + ax'' + f(x') + cx = 0 \tag{2}$$

$$x''' + f(x'') + bx' + cx = 0 \tag{3}$$

$$x_1' = -cx_1+x_2-f(x_1),\ x_2' = -x_1+x_3,\ x_3' = -cx_1+bf(x_1) \tag{4}$$

Let $f(0) = 0$; then each system admits the zero solution the stability properties of which will be examined. In cases (1), (3), and (4) this problem is solved in [4]; equation (2) is added for the sake of completeness. Kasprzyk also investigates the system

$$x_1' = x_2-f(x_1),\ x_2' = -x_1+x_3,\ x_3' = -ax_1 \quad (a > 0) \tag{5}$$

which is, however, equivalent to a special form of equation (3). Introducing

$$x = a^{-2}(x_3 - ax_2) \quad [x' = -a^{-1}x_3,\ x'' = x_1]$$

we derive from (5):

$$x''' + f(x'') + x' + ax = 0$$

Let us mention the generalized Hurwitz conditions of the systems (1) - (4):

$$0 < x^{-1}f(x) < ab \quad (x \neq 0) \tag{1'}$$

$$y^{-1}f(y) > a^{-1}c \quad (y \neq 0) \tag{2'}$$

$$z^{-1}f(z) > b^{-1}c \quad (z \neq 0) \tag{3'}$$

$$0 < x_1^{-1}f(x_1) < b^{-1}c \quad (x_1 \neq 0) \tag{4'}$$

The following theorem is proved by Kasprzyk:

Assume in (1) that $f(x) \in C^1(R)$, $xf(x) > 0$ $(x \neq 0)$,

$$\int_0^\infty \mathrm{Min}[f(s),-f(-s)]ds = +\infty,\ 0 < f'(0) < ab, \text{ and}$$

$|f'(x)| \leq ab$ for all $x \in R$. Then $x(t) \equiv 0$ is a globally asymptotically stable solution of (1).

Assume in (3) that $f(z) \in C^1(R)$, $f'(0) > b^{-1}c$ but $f'(z) \geq b^{-1}c$ for all $z \neq 0$; then $x(t) \equiv 0$ is a globally asymptotically stable solution of (3).

Assume in (4) that $c^2 > b$, $f(x_1) \in C^1(R)$, $0 < f'(0) < b^{-1}c$ but $f'(x_1) \geq 0$ for all $x_1 \neq 0$, $x_1^{-1}f(x_1) < b^{-1}c$. Then the trivial solution of (4) is globally asymptotically stable.

In generalizing this theorem we emphasize the difference between systems (1), (2) and (3), (4).

THEOREM A. Assume in equations (1) and (2) that $f \in C^o(R)$, and that the initial value problem has a uniquely determined solution which is continuously depending upon the initial values. Assume in equation (1) that $0 < x^{-1}f(x) \leq ab$ for all $x \neq 0$ but $0 \in C\ell\{x \neq 0 : x^{-1}f(x) < ab\}$. Then $x(t) \equiv 0$ is a globally asymptotically stable solution of (1). Assume in equation (2) that $y^{-1}f(y) \geq a^{-1}c$ for all $y \neq 0$ but $0 \in C\ell\{y \neq 0 : y^{-1}f(y) > a^{-1}c\}$. Then $x(t) \equiv 0$ is a globally asymptotically stable solution of (2).

THEOREM B. Assume in systems (3) and (4) that $f \in C^1(R)$. Assume in equation (3) that $z^{-1}f(z) > b^{-1}c$ for all $z \neq 0$ and $f'(z) \geq b^{-1}c$ for all z. Then $x(t) \equiv 0$ is a globally asymptotically stable solution of (3). Assume in system (4) that $0 < x_1^{-1}f(x_1) < b^{-1}c$ for all $x_1 \neq 0$ and $f'(x_1) \geq 0$ for all x_1. Then the trivial solution of (4) is globally asymptotically stable.

These theorems will be proved in the following way. Let us consider an autonomous differential equation

$$\underline{x}' = \underline{F}(\underline{x}) \quad [\underline{x} \in R^n,\ \underline{F}(\underline{x}) \in C^o(R^n, R^n)] \tag{6}$$

the solution of which is assumed to be a continuous function of its initial value $\underline{x}(0) = \underline{x}_o$. Let a Liapunov function $V(\underline{x}) \in C^1(R^n, R)$ such that

$$V(\underline{x}) \geq 0,\ V'(\underline{x}) = (\text{grad } V, \underline{F}) \leq 0 \text{ for all } \underline{x} \in R^n$$

be given. Assume that all solutions of (6) are bounded for $t \geq 0$; this is ensured when $V(\underline{x})$ is radially unbounded;

$$V(\underline{x}) \to \infty \text{ as } |\underline{x}| \to \infty$$

Let $E = \{\underline{x} \in R^n : V'(\underline{x}) = 0\}$, and let $M \subset E$ be the maximum invariant subset. Then $d(\underline{x}(t),M) \to 0$ as $t \to \infty$ where $\underline{x}(t)$ is an arbitrary solution of (6). This variant of LaSalle's invariance principle (see [16]) is an immediate consequence of the fact that the bounded solution $\underline{x}(t)$ has a non-empty ω-limit set $\Omega(\underline{x}_0)$ which is an invariant subset of E, i.e. $\Omega(\underline{x}_0) \subset M \subset E$.

Let us further assume that $\underline{F}(0) = 0$ and that the zero solution of (6) is weakly stable. This is ensured by virtue of the principal theorem of Liapunov when $V(\underline{x})$ positive definite. Then we have global asymptotic stability (i.e. weak stability and global attractiveness of the equilibrium point $\underline{x} = 0$) when $M = \{0\}$. But this is also true when $\underline{x}(t) \to 0$ as $t \to \infty$ provided that $\underline{x}(0) \in M$. Since all ω-limit trajectories are contained in M, the origin is an ω-limit point of each solution; being a stable equilibrium point, it is the only ω-limit point.

PROOF OF THEOREM A

Introducing the variables

$$x_1 = x,\ x_2 = ax' + x'',\ x_3 = x'$$

we transform equation (1) into the system

$$x_1' = x_3,\ x_2' = -f(x_1) - bx_3,\ x_3' = x_2 - ax_3 \tag{7}$$

The function

$$V(\underline{x}) = (bx_1 + x_2)^2 + (x_2 - ax_3)^2 + bx_3^2 + 2a\int_0^{x_1} f(u)\,du$$

where $\underline{x} = \mathrm{col}(x_1,x_2,x_3)$, is positive definite and radially unbounded. Its total derivative, by virtue of system (7), is

$$V' = -2a^{-1}(abx_1 - f(x_1))f(x_1) - 2a^{-1}(f(x_1)+ax_2-a^2x_3)^2 \leq 0$$

Consequently it is a sufficient criterion for weak stability of the origin as well as for boundedness of all solutions. In order to apply the invariance principle we investigate those solutions $\underline{x}(t)$ for which $V'(\underline{x}(t)) \equiv 0$. That means:

$$abx_1(t) \equiv f(x_1(t)),\ f(x_1(t)) \equiv -ax_2(t) + a^2x_3(t)$$

The components x_2, x_3 are solutions of the linear system

$$x_2' = ax_2 - (a^2+b)x_3,\ x_3' = x_2 - ax_3 \tag{8}$$

from which we derive $x_i'' + bx_i = 0$ $(i = 2,3)$. Therefore,

$$x_1(t) = -b^{-1}x_2(t) + ab^{-1}x_3(t) = p\sin(\sqrt{b}\,t + \phi),\ p \geq 0$$

In case $p > 0$ we have $f(x_1) = abx_1$ for all $x_1 \in [-p,+p]$ which is in contradiction to the assumption. So, we obtain $x_1(t) \equiv 0$ and, according to (7), $x_2(t) = x_3(t) \equiv 0$. The maximum invariant subset $M \subset E$ reduces to the origin.

By means of

$$x_1 = x',\ x_2 = ax' + x'',\ x_3 = x$$

equation (2) is transformed into the system

$$x_1' = -ax_1 + x_2,\ x_2' = -f(x_1) - cx_3,\ x_3' = x_1 \tag{9}$$

The positive semidefinite function

$$V(\underline{x}) = (2a^2)^{-1}c(x_1 + ax_3)^2 + (2a)^{-1}x_2^2$$

$$+ a^{-1}\int_0^{x_1}(f(u) - a^{-1}cu)du$$

has the total derivative, by virtue of (9),

$$V' = -x_1(f(x_1) - a^{-1}cx_1) \leq 0$$

This time Liapunov's principal theorem on stability is not applicable but it can be replaced by a simple argument. Let $\underline{x}(0) = \underline{x}_o$, $V(\underline{x}_o) = V_o$; since $V(\underline{x}(t))$ is monotone-decreasing, we can estimate for all $t \geq 0$:

$$|x_1(t) + ax_3(t)| \leq (2a^2c^{-1}V_o)^{1/2} \to 0 \text{ as } |\underline{x}_o| \to 0$$

$$|x_2(t)| \leq (2aV_o)^{1/2} \to 0 \text{ as } |\underline{x}_o| \to 0$$

The differential equation for the first component $x_1(t)$ is considered as a nonhomogeneous linear one ($x_2(t)$ being a bounded forcing term):

$$x_1(t) = x_1(0)e^{-at} + \int_0^t e^{-a(t-s)}x_2(s)\,ds$$

hence

$$|x_1(t)| \leq |x_1(0)| + (2a^{-1}V_o)^{1/2} \to 0 \text{ as } |\underline{x}_o| \to 0$$

Summarizing, we have boundedness of all solutions and weak stability of the zero solution.

Finally, let us show that $V'(\underline{x}(t)) \equiv 0$ implies that $\underline{x}(t) \to 0$ as $t \to \infty$. Obviously, we have

$$f(x_1(t)) \equiv a^{-1}cx_1(t)$$

Taking account of this equation we obtain linear differential equations for the considered solutions (belonging to M); the differential system is equivalent to the third order equation

$$x''' + ax'' + a^{-1}cx' + cx = 0$$

the characteristic polynomial of which has the roots $-a$, $\pm i\sqrt{b}$ (where the abbreviation $b = a^{-1}c$ is used). Thus, $x_1 \equiv x'$ can be represented as

$$x_1(t) = p \sin(\sqrt{b}t + \phi) + qe^{-at}, \; p \geq 0$$

from which we conclude that $f(x_1) = a^{-1}cx_1$ for all $x_1 \in [-p,+p]$, and $p = 0$, by assumption. Then $x_1(t) = qe^{-at}$, $x_2(t) \equiv 0$, $x_3(t) = -qa^{-1}e^{-at}$ (but not necessarily $q = 0$). Since all trajectories belonging to M are approaching the origin, it must be a globally attractive equilibrium point.

Note. Systems (7) and (9) are of Tuzov type. Tuzov [17] studied autonomous third order systems including one nonlinear term which depends upon one of the variables, but which does not occur in the differential equation for this variable. For a detailed discussion see [15]. Such a system is called of Pliss type when the nonlinear term is contained in the differential equation for its argument. Systems (3) and (4) are two examples. Whereas Aizerman's conjecture is valid for the system of Tuzov type (apart from one exceptional subcase), the situation concerning the system of Pliss type is much more complicated. In studying the asymptotic behavior of this system (see [10],[12], [13],[15]), there are numerous subcases to be distinguished. Aizerman's conjecture fails in several subcases.

PROOF OF THEOREM B

Equation (3) is equivalent to the system

$$x_1' = x_2 - f(x_1),\ x_2' = -bx_1 - cx_3,\ x_3' = x_1 \tag{10}$$

which is obtained by means of the transformation

$$x_1 = x'',\ x_2 = -cx - bx',\ x_3 = x'$$

Consider the positive definite function $V(\underline{x})$,

$$2V = b(b^{-1}cx_1 - f(x_1) + x_2)^2 + (bx_1 + cx_3)^2 +$$

$$+ 2c\int_0^{x_1}(f'(u) - b^{-1}c)udu$$

possessing the total derivative, by virtue of (10),

$$V' = -b(f'(x_1) - b^{-1}c)(f(x_1) - x_2)^2 \leq 0$$

Boundedness of the solutions and weak stability of the zero solution is proved as in the previous case:

$$|b^{-1}cx_1(t)-f(x_1(t))+x_2(t)| \leq (2b^{-1}V_o)^{1/2}$$

$$|bx_1(t)+cx_3(t)| \leq (2V_o)^{1/2} \quad (t \geq 0)$$

$$x_1' = b^{-1}cx_1 + (b^{-1}cx_1-f(x_1)+x_2), \text{ i.e.}$$

$$|x_1(t)| \leq |x_1(0)| + c^{-1}(2bV_o)^{1/2} \quad (t \geq 0)$$

Let $\underline{x}(t)$ be a solution satisfying the additional condition $V'(\underline{x}(t)) \equiv 0$ (which means that $\underline{x}(t) \in M$). (a) Let $f'(x_1(t_o)) > b^{-1}c$ for some t_o. Then there is an interval $i = [t_o, t_o+h]$ where $f(x_1(t)) = x_2(t)$ and $x_1'(t) = x_2'(t) \equiv 0$. We derive successively from (10) that for all $t \in i$, $x_3'(t) = c^{-1}(x_2''(t) + bx_1'(t)) = 0$, i.e. $x_1(t) = 0$ and $x_2(t) = 0$, $x_3(t) = 0$. By virtue of uniqueness, the considered solution is the trivial one. (b) Let $f'(x_1(t)) \equiv a$ for all $t \in R$ $(a = b^{-1}c)$; from (10) it follows that $x_1(t) \in C^3(R)$ and $x_1''' + ax_1'' + bx_1' + cx_1 = 0$,

$x_1(t) = p\sin(\sqrt{b}t + \phi) + qe^{-at}$ $(p \geq 0)$. If $p > 0$ or $p = 0$, $q \neq 0$ we obtain a contradiction to the assumption: $f(x_1) = b^{-1}cx_1$ on the segment of the real axis with extremities $-p$, $+p$ or 0, q, respectively. Consequently, $x_1(t) \equiv 0(\underline{x}(t) \equiv 0)$. Summarizing (a) and (b) we have that $M = \{0\}$.

Note. Global asymptotic stability of the trivial solution can also be proved under the conditions $f(z) \in C^0(R)$, $b^{-1}c \leq z^{-1}f(z) \leq b^{-1}c + c^{-1}b^2$ for all $z \neq 0$, but $0 \in C\ell\{z \neq 0 : z^{-1}f(z) > b^{-1}c\}$. A nonlinear function $f(z)$ satisfying the conditions

$$z^{-1}f(z) > b^{-1}c \ (z \neq 0), \ f'(0) > b^{-1}c + c^{-1}b^2$$

has been constructed by Pliss [11] (see [15]) in such a way that there are nontrivial periodic solutions. Hence, Aizerman's conjecture fails in the case of equation (3).

In case of system (4) we consider the positive definite function $V(\underline{x})$,

$$2V = ((1+b)f(x_1) - x_2)^2 + (x_1 - x_3)^2$$

$$+ 2(1+b)\int_0^{x_1} f'(u)(cu - bf(u))du$$

yielding the total derivative, by virtue of (4),

$$V' = -(1+b)f'(x_1)(f(x_1) + cx_1 - x_2)^2 \leq 0$$

V is a sufficient criterion for weak stability of the zero solution. A further consequence of the monotonous decrease of V along every solution is the boundedness of the terms

$$(1+b)f(x_1(t)) - x_2(t), \ x_1(t) - x_3(t) \text{ for } t \geq 0$$

If $|x_1(t) - x_3(t)| \leq X$ then $\operatorname{sgn} x_1(t) = \operatorname{sgn} x_3(t)$ in case $|x_3(t)| > X$, hence

$$|x_3|' = x_3' \operatorname{sgn} x_1 = -b|x_1|(b^{-1}c - x_1^{-1}f(x_1)) < 0$$

and

$$|x_3(t)| \leq \operatorname{Max}(|x_3(0)|, X) \text{ for } t \geq 0$$

Now we conclude that the components $x_1(t)$, $x_2(t)$ are bounded also.

Finally, let us show again that the maximum invariant subset $M \subset E$ reduces to the origin. If $\{\underline{x}(t) : -\infty < t < +\infty\} \subset M$ then either (a) $f'(x_1(t_0)) > 0$ for some t_0 or (b) $f'(x_1(t)) \equiv 0$. (a) There is an interval $i = [t_0, t_0 + h]$ where $x_1' = -cx_1 + x_2 - f(x_1) = 0$ (hence $x_2' = x_1'' = 0$), $x_1 = x_3$ (hence $x_3' = 0$, i.e. $cx_1 = bf(x_1)$ which means that $x_1 = 0$ and $x_2 = x_3 = 0$). The considered solution is the trivial one. (b) $x_1(t) \in C^3(R)$, $f(x_1(t)) \equiv f_o$ (constant) and $x_1''' + cx_1'' + x_1' + cx_1 = bf_o$. Hence $x_1(t) = p \sin(t + \phi) + qe^{-ct} + c^{-1}bf_o$ ($p \geq 0$). Consider the sequence $\{t_k\}$, $t_k = k\pi - \phi$; $x_1(t_k) \to u = c^{-1}bf_o$, $f(x_1(t_k)) = f_o = f(u)$ and $cu - bf(u) = 0$, i.e. $u = 0$ ($f_o = 0$). An immediate consequence is that $x_1(t) \equiv 0$ (since $x_1 f(x_1)$ is positive definite) and $x_2(t) = x_3(t) \equiv 0$.

REFERENCES

1. J. O. C. Ezeilo, A stability result for the solutions of certain third order differential equations, J. London Math. Soc. 37 (1962), 405-409.
2. J. O. C. Ezeilo, A stability result for a certain third order differential equation, Ann. Mat. Pura Appl. 72 (1966), 1-9.
3. P. Hartman and C. Olech, On global asymptotic stability of solutions of differential equations, Trans. Amer. Math. Soc. 104 (1962), 154-178.
4. S. Kasprzyk, Stability in the large of certain nonlinear systems of differential equations of order three, Ann. Polon. Math. 25 (1972), 241-247.
5. J. P. LaSalle, Liapunov's second method, in Stability Problems of Solutions of Differential Equations, Proceedings of NATO Advanced Study Institute, Padua, Italy, Edizioni "Oderisi", Gubbio, 1966, 95-106.
6. J. P. LaSalle, An invariance principle in the theory of stability, in Differential Equations and Dynamical Systems, Proceedings of an International Symposium, Puerto Rico, Academic, New York, 1967, 277-286.
7. J. P. LaSalle, Stability theory for ordinary differential equations, J. Differential Equations 4 (1968), 57-65.
8. J. P. LaSalle, Stability theory and invariance principles, in Dynamical Systems, An International Symposium, Vol. 1, Academic, New York, 1976, 211-222.
9. J. P. LaSalle and S. Lefschetz, Stability by Liapunov's Direct Method, Academic, New York, 1961.
10. W. Müller, Sufficient conditions for the boundedness of solutions of a system of three differential equations (German), Abh. Deutsch. Akad. Wiss, Berlin Kl. Math. Phys. Tech., 1967, Nr. 2.
11. V. A. Pliss, Investigation of a nonlinear differential equation of order three (Russian), Dokl. Akad. Nauk SSSR 111 (1956), 1178-1180.

12. V. A. Pliss, Some Problems Concerning the Theory of Global Stability of a Motion (Russian), Leningrad, 1958.

13. V. A. Pliss, Nonlocal Problems of the Theory of Oscillations, Academic, New York, 1966.

14. R. Reissig, Global asymptotic stability of certain nonlinear autonomous differential equations, Atti Accad. Naz. Lincei Rend. Cl. Sci. Fis. Mat. Natur. (8) 53 (1973), 39-45.

15. R. Reissig, G. Sansone, and R. Conti, Non-linear Differential Equations of Higher Order, Noordhoff, Leyden, 1974.

16. N. Rouche and J. Mawhin, Ordinary Differential Equations, Vol. 2: Stability and Periodic Solutions (French), Paris, 1973.

17. A. P. Tuzov, On the global stability of a control system (Russian), Vestnik Leningrad. Univ. 1 (1957), 57-75.

Chapter 16

A REPRESENTATION FOR INPUT-OUTPUT OPERATORS FOR HEREDITARY SYSTEMS

JAMES A. RENEKE
Department of Mathematics
Clemson University
Clemson, South Carolina

Suppose that S is an interval of numbers containing 0, X is a linear space, and G is a linear space of functions from S into X. Suppose that N is a function from S into the class of pseudonorms on G such that (1) if u is in S and f is in G then $N_u(f) = 0$ if and only if $f(x) = 0$ (the zero of X) for each x in S which does not exceed u, (2) if [u,v] is a subinterval of S and f is in G then $N_u(f) \leq N_v(f)$, and (3) {G,N} is complete. Suppose that H denotes the class of linear transformations of G, 1 denotes the identity in H, and P is a function from S into the indempotents of H such that $N_v(P_u f) = N_u(f)$ for each subinterval [u,v] of S and f in G. We will let A denote the subset of H to which A belongs if and only if (1) $[Af](u) = f(u)$ for f in G and u in S not exceeding 0, and (2) there is a nondecreasing function k from S to the numbers such that

$$N_w(P_v[1-A]f - P_u[1-A]f) \leq (L)\int_u^v N_x(f)\,dk(x) \qquad (*)$$

for each (f,g) in G x G and nondecreasing sequence {u,v,w} in

S. We will define a linear hereditary system to be in a 6-tuple {S,X,G,N,P,A}, where S,X,G,N, and P are as above and A is in A, and call A the input-output operator for the system.

Our purpose in this chapter is to obtain a representation for a linear input-output operator and to develop a stability result for the system in terms of that representation.

In what follows we will take $S = [0,\infty)$, d to be a positive integer, X to be the space of d-tuples of complex numbers with the Euclidean inner product $\langle\cdot,\cdot\rangle$ and norm $|\cdot| = \langle\cdot,\cdot\rangle^{1/2}$, G to be the functions from S into X which have bounded variation on compact intervals, $N_u(f) = |f(0)| + \int_0^u |df|$ for each f in G and u in S,

$$[P_u f](x) = \begin{cases} f(x), & 0 \le x \le u \\ f(u), & u \le x \end{cases}$$

for each f in G and u in S, A to be a member of A, and k to be an increasing function from S to the numbers such that the pair (A,k) satisfies condition (*). We will obtain a representation of A on a subspace of G as a 'matrix' using the theory of reproducing kernels for inner product spaces. Our inner product space will be those members of G which are Hellinger integrable [1] with respect to k. Since the definition and properties of the Hellinger integral are not so widely known, we will summarize a few facts here.

A member f of G is said to be Hellinger integrable with respect to k on a subinterval [u,v] of S provided there is a number b such that

$$\sum_{p=1}^{n} |df(s_{p-1},s_p)|^2/dk(s_{p-1},s_p) \le b$$

for each partition $\{s_p\}_0^n$ of [u,v]. If each of f and g is Hellinger integrable on [u,v] then there is a complex number J with the property that for each positive number c there is a partition $\{s_p\}_0^n$ of [u,v] such that

$$|J - \sum_{q=1}^{m} \langle df(t_{q-1},t_q),\ dg(t_{q-1},t_q)\rangle / dk(t_{q-1},t_q)| < c$$

for each refinement $\{t_q\}_0^m$ of s. We will denote such a number J by $\int_u^v \langle df,\ dg\rangle/dk$.

For each positive number u let $\overline{G}_u$ denote the subset of G to which f belongs if and only if f is Hellinger integrable on $[0,u]$ with respect to k. Let $\overline{Q}_u$ denote the pseudo-inner product for $\overline{G}_u$ defined as follows:

$$\overline{Q}_u(f,g) = \langle f(0),g(0)\rangle + \int_0^u \langle df,dg\rangle/dk$$

for each (f,g) in $\overline{G}_u \times \overline{G}_u$. We will take $\overline{N}_u = \overline{Q}_u(\cdot,\cdot)^{1/2}$. If $\overline{G}_\infty$ is the common part of the $\overline{G}_u$'s then $\{\overline{G}_\infty,\overline{N}\}$ is complete.

Let K denote the function from S x S into $\mathcal{H}$ defined by

$$K(u,v)x = \begin{cases} (k(u)-k(0)+1)x, & 0 \le u \le v \\ (k(v)-k(0)+1)x, & v \le u \end{cases}$$

for each (u,v) in S x S and x in X. K is a reproducing kernel for $\{\overline{G}_\infty,\overline{Q}\}$ in the sense that (1) $K[\ ,u]x$ is in $\overline{G}_\infty$ for each u in S and x in X, and (2) if $[u,v]$ is a subinterval of S, x is in X, and f is in G, then $\overline{Q}_v(f,K[\ ,u]x) = \langle f(u),x\rangle$.

THEOREM 1. The restriction of A to $\overline{G}_\infty$ is a reversible function from $\overline{G}_\infty$ onto $\overline{G}_\infty$ and is continuous with respect to $\overline{N}$.

Proof. We have shown [2] that A is a reversible function from G onto G. Suppose that f is in G and u is a positive number. If $\{s_p\}_0^n$ is a partition of $[0,u]$ then

$$\sum_{p=1}^{n} |[(1-A)f](s_p) - [(1-A)f](s_{p-1})|^2/dk(s_{p-1},s_p)$$

$$\le \sum_{p=1}^{n} N_u^2(P_{s(p)}[1-A]f - P_{s(p-1)}[1-A]f)/dk(s_{p-1},s_p)$$

$$\leq \sum_{p=1}^{n} [(L)\int_{s(p-1)}^{s(p)} N_x(f)\,dk(x)]^2/dk(s_{p-1},s_p)$$

$$\leq N_u^2(f)\,dk(0,u)$$

Therefore 1-A maps G into $\overline{G}_\infty$ and so A maps $\overline{G}_\infty$ into $\overline{G}_\infty$. Suppose that h is in $\overline{G}_\infty$. Then $A^{-1}h$ is in G and $[1-A]A^{-1}h = A^{-1}h - h$ is in $\overline{G}_\infty$. Therefore $A^{-1}h$ is in $\overline{G}_\infty$ or A maps $\overline{G}_\infty$ onto $\overline{G}_\infty$.

The above inequality shows that

$$\begin{aligned}\overline{N}_u([1-A]f) &\leq N_u(f)\,dk^{1/2}(0,u)\\ &\leq \overline{N}_u(f)\,dk^{1/2}(0,u)(1+dk^{1/2}(0,u))\end{aligned}$$

Therefore the restriction of A to $\overline{G}_\infty$ is continuous with respect to $\overline{N}$.

The following theorem appears in MacNerney's paper [1] in a more general setting.

THEOREM 2. There is a function L from S x S into the linear transformations of X such that, for each u in S and x in X, (1) L[,u]x is in $\overline{G}_\infty$, and (2) $\langle [Af](u),x\rangle = \overline{Q}_u(f,L[\ ,u]x)$ for each f in $\overline{G}_\infty$.

Proof. For each (u,x) in S x X, let L[,u]x be that number g of $\overline{G}_\infty$ such that $\langle [Af](u),x\rangle = Q_u(f,g)$ for each f in $\overline{Q}_u$ and $g(v) = g(u)$ for $u \leq v$. If [u,v] is a subinterval of S and (x,y) is in X x X then $\langle x,L(u,v)y\rangle = Q_v(K[\ ,u]x,L[\ ,v]y) = \langle [A(K[\ ,u]x)](v),y\rangle$. Therefore L is a function from S x S into the continuous linear transformations of X.

We will assume for the remainder of the chapter that k is continuous and k(0) = 0. For each positive number b and f in G such that f(0) = 0, we will denote by $S_b(f)$ and $S_{-b}(f)$ the members of G defined, respectively, as follows:

$$[S_b f] = \begin{cases} 0, & 0 \leq u \leq b \\ f(k^{-1}(k(u)+b)), & b \leq u \end{cases}$$

and $[S_{-b}f](u) = f(k^{-1}(k(u)+b))$. We will say that the hereditary system with input-output operator A is time invariant provided, for each positive number b and f in G such that $f(0) = 0$, $Af = S_{-b}AS_b f$.

THEOREM 3. If the system is time invariant, then $L(v,w) - L(u,w) = L(k^{-1}(k(v)+b),k^{-1}(k(w)+b)) - L(k^{-1}(k(u)+b),k^{-1}(k(w)+b))$ for each positive number b and $0 \le u \le v \le w$.

Proof. If (x,y) is in X x X then $<y,L(v,w)x-L(u,w)x> = <[A(K[\ ,v)y-K[\ ,u]y)](w),x> = <[AS_b(K[\ ,v]y-K[\ ,u]y)](k^{-1}(k(w)+b)),x>$. If $0 \le t \le b$ then $0 = S_b(K[\ ,v]y-K[\ ,u]y)(t) = K(t,k^{-1}(k(v)+b))y - K(t,k^{-1}(k(u)+b)y$. If $b \le t$ then

$$
\begin{aligned}
&S_b(K[\ ,v]y-K[\ ,u]y)(t)\\
&= K(k^{-1}(k(t)-b),v)y - K(k^{-1}(k(t)-b),u)y\\
&= \begin{cases} 0, & k^{-1}(k(t)-b)\le u \text{ or } t\le k^{-1}(k(u)+b)\\ k(t)-b-k(u), & \begin{cases} u\le k^{-1}(k(t)-b)\le v \\ \text{or} \\ k^{-1}(k(u)+b)\le t\le k^{-1}(k(v)+b)\end{cases}\\ k(v)-k(u), & v\le k^{-1}(k(t)-b) \text{ or } k^{-1}(kw)+b)\le t\end{cases}\\
&= K(t,k^{-1}(k(v)+b))y - K(t,k^{-1}(k(u)+b))y.
\end{aligned}
$$

Therefore $<y,L(v,w)x-L(u,w)x> = <[A(K[\ ,k^{-1}(k(v)+b)]y - K[\ ,k^{-1}(k(u)+b]y)](k^{-1}(k(w)+b)),x> = <y,L(k^{-1}(k(v)+b),k^{-1}(k(w)+b))x - L(k^{-1}(k(w)+b),k^{-1}(k(w)+b))x>$.

The system is said to be BIBO stable provided: if f is in $\overline{G}_\infty$ and f(S) is bounded, then [Af](S) is bounded.

THEOREM 4. If $\int_0^\infty |dL[u,I]| < \infty$ for each u in S and $\limsup_{h \to 0+} (1/k(h))\int_0^\infty |d\{L[0,I] - L(k^{-1}(h),I]\}| < \infty$ then the system is BIBO stable.

Proof. Suppose that f is in $\overline{G}_\infty$, m is a positive number, $|f(u)| \le m$ for each u in S, and x is in X. If u is in S and c is a positive number, then there is a positive integer n such that if $\{s_p\}_0^n$ is the partition of [0,u] given by $s_0 = 0$ and $s_p = k^{-1}(k(s_{p-1}) + k(u)/n)$ for $p = 1,2,\ldots,n$, then $|<[Af](u),x>| = |Q_u(f,L[\ ,u)x)| \le c + |<f(0),L(s_0,u)x> + \sum_{p=1}^n <df(s_{p-1},s_p), L(s_p,u)x - L(s_{p-1},u)x>/dk(s_{p-1},s_p)| = c + |<f(s_0),L(s_0,h)x>$

$+ \sum_{p=1}^{n} \langle f(s_p), L(s_1,s_{n-p+1})x - L(s_0,s_{n-p+1})x - L(s_1,s_{n-p})x + L(s_0,s_{n-p})x\rangle\cdot(n/k(u)) - \langle f(s_0), L(s_1,u)x - L(s_0,u)x\rangle\cdot(n/k(u))|$

$\leq c + m(1+\int_0^\infty |dL[0,I]|)|x| + 2m|x| \{\limsup_{h\to 0+} (1/k(h))\int_0^\infty |d\{L[0,I] - L[k(h),I]\}| + c.$ Therefore $|[Af](u)| \leq m(1+\int_0^\infty |dL[0,I]| + 2\limsup_{h\to 0+} (1/k(h))\int_0^\infty |d\{L[0,I] - L[k(h),I]\}|)$, for each u in S, and so the system is BIBO stable.

The significance of the result lies in the fact that we can deduce BIBO stability from a 'row' condition on the matrix L. In other words, we can achieve BIBO stability by imposing conditions on the class of functions A(K[,u]x) where (u,x) is in S x X.

REFERENCES

1. J. S. MacNerney, Hellinger integrals in inner product spaces, J. Elisha Mitchell Sci. Soc. 76 (1960), 252-273.
2. J. A. Reneke, The input-output functions for a class of hereditary systems, submitted for publication.
3. J. C. Williams, The generation of Lyapunov functions for input-output stable systems, SIAM J. Control 9 (1971), 105-134.

Chapter 17

STABILITY OF PASSIVE AMCD-SPACECRAFT EQUATIONS OF MOTION: A FUTURE GENERATION SPACECRAFT ATTITUDE CONTROL SYSTEM

A. S. C. SINHA
Departments of Electrical Engineering and Mathematics
Indiana Institute of Technology
Fort Wayne, Indiana

INTRODUCTION

The system under consideration represents a new concept in attitude control actuators. Basically, it consists of an annular rim driven by a rim drive motor and supported by magnetic bearing. Detailed descriptions and advantages are given by Anderson and Groom [1].

Accurate modeling is extremely difficult to achieve for flexible dissipative spacecraft structures and various damping mechanisms that have been used in the spacecraft. The energy dissipation in either the high speed Annular Momentum Control Device (AMCD) or the despun main spacecraft structure introduced through a damping mechanism, is modeled by one ball-in-tube damper. The AMCD is considered symmetrical and the energy dissipation must, therefore, be symmetric in the two transverse axes. Three additional particles, each of whose mass is equal to the total mass of damping mechanism, are rigidly attached to the spacecraft in such a way that the combination of damper and three particles becomes inertially symmetrical about the spin axis.

Equations of motion of this system are derived here in rather more general terms, in order to permit easier visualization of the more general case. A new mathematical model is introduced and analyzed from the first principles. Euler's equations for the spacecraft and AMCD rim are written, where external disturbance torques are included. The torque on the AMCD rim about the axis of rotation resulting from magnetic bearing torques are expressed as a function of the rotation and the axial magnetic bearing constants. The torques experienced due to three balancing masses and the damper are also included.

Since the inertial attitude stability of the spin axis is of interest, the natural choice is the set of attitude coordinates (e.g., Euler's angles) defining the orientation of the body frame with respect to the inertial frame. The equations are transformed into inertial frame which on considerable algebraic manipulation gives eight second degree differential equations with periodic coefficients, coupled even in the highest order.

A necessary and sufficient condition for the asymptotic stability of AMCD-spacecraft equation is developed. The criterion is based on a theorem of LaSalle.

It is only recently that dual-spin stabilization technology has been developed and used, although it was long ago that Euler formulated and described the motion of rigid bodies [3,6]. With the increasing complexities of dual-spin spacecraft, the stability and dynamic behaviors of the system with the energy dissipation taken into account have recently become of more concern to system designers. In the last several years, the concept, analysis, and design of dual-spin bodies have been developed by outstanding works of many people; see, for example [1,2,4,5,7,9-11,13-25, and 27].

FORMULATION

Let T_b, I_b, ω_b and H_b be the external torques, the inertia matrix, the body rates of rotation, and the angular momentum respectively for an arbitrary rigid body. This body represents the rotating rim of AMCD. Then the standard Euler's equation of motion has the following form

$$T_b = I_b\dot{\omega}_b + \omega_b H_b \tag{1}$$

Since the magnetic bearings which produce the torques, T_b, are fixed to the spacecraft, it is required to transform equation (1) to a second arbitrarily oriented axes system. The transformation, E_{ab}, is defined by the equation

$$v_a + E_{ab}v_b$$

where v_a and v_b are arbitrary vectors and the transformation is defined by denoting $c\alpha$ = cos α, and $s\alpha$ = sin α, as

$$E_{ab} = \begin{bmatrix} c\alpha & -s\alpha & 0 \\ s\alpha & c\alpha & 0 \\ 0 & 0 & 1 \end{bmatrix} \tag{2}$$

Then the Euler's equation (1) is premultiplied by E_{ab} and noting that $E_{ab}^{-1}\ E_{ab}$ = I (unit matrix), (1) is rewritten as

$$E_{ab}T_b = E_{ab}I_bE_{ab}^{-1}E_{ab}\dot{\omega}_b + E_{ab}\omega_bE_{ab}^{-1}E_{ab}H_b$$

Now define the following:

$$T_a = E_{ab}T_b$$

$$I_a = E_{ab}I_bE_{ab}^{-1}$$

$$\omega_a = E_{ab}\omega_bE_{ab}^{-1} \quad \text{(matrix)}$$

$$\omega_a = E_{ab}\omega_b \quad \text{(vector)} \tag{3}$$

Differentiating (3) yields

$$\dot{\omega}_a = E_{ab}\dot{\omega}_b + \dot{E}_{ab}\omega_b$$

or

$$E_{ab}\dot{\omega}_b = \dot{\omega}_a - \dot{E}_{ab}\omega_b$$

Combining the above equations gives

$$T_a = I_a\dot{\omega}_a + \omega_a H_a - I_a \dot{E}_{ab} E_{ab}^{-1} \omega_a \tag{4}$$

Assuming AMCD rim symmetry, that is, $I_{bx} = I_{by} = I_a$, equation (4) can be expanded as

$$T_{ax} = I_a\dot{\omega}_{ax} + (I_{az}-I_a)\omega_{ay}\omega_{az} + \dot{\alpha} I_a \omega_{ay} \tag{5}$$

$$T_{ay} = I_a\dot{\omega}_{ay} + (I_a-I_{az})\omega_{ax}\omega_{az} - \dot{\alpha} I_a \omega_{ax} \tag{6}$$

$$T_{az} = I_{az}\dot{\omega}_{az} \tag{7}$$

Equations (5) - (7) can now be modified to separate the effect of the large AMCD rim spin velocity by introducing the variable ω^*_{az} which will represent the angular velocity of the a-coordinate system about the z-axis, where $\omega_{az} = \omega^*_{az} + \alpha$. Substituting in equations (5) - (7) gives

$$T_{ax} + G_{ax} = I_a\dot{\omega}_{ax} + (I_{az}-I_z)\omega_{ay}\omega^*_{az} + I_{az}\dot{\alpha}\omega_{ay} \tag{8}$$

$$T_{ay} + G_{ay} = I_a\dot{\omega}_{ay} + (I_a-I_{az})\omega^*_{az}\omega_{ax} - I_{az}\dot{\alpha}\omega_{ax} \tag{9}$$

$$T_{az} + G_{az} = I_{az}\dot{\omega}^*_{az} + I_{az}\ddot{\alpha} \tag{10}$$

where the additional terms G_{ax}, G_{ay}, G_{az} are external disturbance torques, added to (5) - (7) due to three balancing masses and one ball-in-tube damper; these are derived in the next section.

When the AMCD and spacecraft centers-of-mass are coincidence, a second set of Euler's equations for the spacecraft similar to (8) - (10) follows:

$$T_{sx} + G_{sx} = I_x\dot{\omega}_{sx} + (I_{sz}-I_s)\omega_{sy}\omega^*_{sz} + I_{sz}\dot{\beta}\omega_{sy} \tag{11}$$

$$T_{sy} + G_{sy} = I_s\dot{\omega}_{sy} + (I_s-I_{az})\omega_s\omega^*_{sx} - I_{sz}\dot{\beta}\omega_{sx} \tag{12}$$

$$T_{sz} + G_{sz} = I_{sz}\dot{\omega}^*_{sz} + I_{sz}\ddot{\beta} \tag{13}$$

The rates ω^*_{az} and ω^*_{sz} will be set equal to zero, and the a- and s-coordinate axes assumed nearly coincident except for small transverse relative rotations. This allows a simplification in the calculation of the interaction torques by introducing a transformation matrix from the s-coordinate system to the a-coordinate system as follows: $E_{as} = E_{ai}E_{is} = E_2(\theta_a)E_1(\phi_a)$ $[E_2(\theta_s)E_1(\phi_s)]^{-1} = E_2(\theta_a)E_1(\phi_a)E_1^{-1}(\phi_s)E_2^{-1}(\theta_s) = E_2(\theta_a)E_1(\phi_a)$ $E_1(-\phi_s)E_2(-\theta_s) = E_2(\theta_a)E_1(\phi_a-\theta_s)E_2(-\theta_s)$ where the subscript i is introduced to represent an inertial reference and the notation E_j(arg.), j = 1 and 2, refers to a transformation from one coordinate system to another which has been rotated through an arg about an axis j. The Euler angles chosen to represent the position of the a- and s-coordinate axes are ϕ and θ with a 1-2 rotation sequence selected.

By expanding the matrices in the last equation for the case when $\phi_a - \phi_s$ and $\theta_a - \theta_s$ are small, the following is found to hold: $E_{as} \doteq E_2(\theta_a-\theta_s)E_1(\phi_a-\phi_s) \doteq E_1(\phi_a-\phi_s)E_2(\theta_a-\theta_s)$. This is true irrespective of the magnitudes of one set of variables (ϕ_a,θ_a) or (ϕ_s,θ_s) and corresponds to a single rotation of the a-coordinate system about an axis in the x-y plane of s-coordinate system. This rotation represents the physical rotation of the plane of the AMCD rim with respect to the spacecraft.

With the rates ω^*_{az} and ω^*_{sz} set equal to zero and the a- and s-coordinate axes nearly coincident except for small transverse relative rotation, equations (8) - (13) are rewritten in the form:

$$T_{ax} + G_{ax} = I_a\ddot{\phi}_a + H_a\dot{\theta}_a \tag{14}$$

$$T_{ay} + G_{ay} = I_a\ddot{\theta}_a - H_a\dot{\phi}_a \tag{15}$$

$$T_{az} + G_{az} = \dot{H}_a \tag{16}$$

$$T_{sx} + G_{sx} = I_s\ddot{\phi}_s + H_s\dot{\theta}_s \tag{17}$$

$$T_{sy} + G_{sy} = I_s\ddot{\theta}_s - H_s\dot{\phi}_s \tag{18}$$

$$T_{sz} + G_{sz} = \dot{H}_s \tag{19}$$

$$H_a = I_{az}\dot{\alpha} \tag{20}$$

$$H_s = I_{sz}\dot{\beta} \tag{21}$$

where G is damper torque vector and torque due to balancing masses. T is bearing and spin torque vectors.

Equations (14) - (21) represent equations of motion for AMCD-spacecraft and will be further analyzed. In the next section, the torques G and T are developed.

TORQUE EQUATION

The torques G_{ax}, G_{ay}, and G_{az} due to three balancing masses and the damper system will now be derived by kinematic principle. Similarly, the torques G_{sx}, G_{sy}, and G_{sz} for the spacecraft will be introduced omitting the algebra. Then, the bearing and spin torque vectors T are derived.

By kinematic principle $\bar{v} = \dot{\bar{r}} + \bar{\omega} \times \bar{r}$ where "·" denotes differentiation with respect to body axis. Then, the acceleration of the mass $\bar{a}$ is $\bar{a} = \dot{\bar{v}} + \bar{\omega} \times \bar{v} = \ddot{\bar{r}} + \dot{\bar{\omega}} \times \bar{r} + 2\bar{\omega} \times \dot{\bar{r}} + \bar{\omega} \times (\bar{\omega} \times \bar{r})$. The inertial force $\bar{F}$ due to this acceleration is

$$\begin{aligned} \bar{F} &= -m\bar{a} \\ &= -m(\ddot{\bar{r}} + \dot{\bar{\omega}} \times \bar{r} + 2\bar{\omega} \times \dot{\bar{r}} + \bar{\omega} \times (\bar{\omega} \times \bar{r})) \end{aligned} \tag{22}$$

The torque is then given by $\bar{G}_b = -\bar{F} \times \bar{r}$; $\bar{r} = \bar{r}_0 + \bar{P}$. With the damper

$$r_{ox_1} = a,\ r_{oy_1} = 0,\ r_{oz_1} = 0;\ P_x = P_y = 0,\ P_z = P \tag{23}$$

(the motion of the damper is constrained to vibrate along the z-axis only) and balancing weights

$$r_{ox_2} = 0,\ r_{oy_2} = a,\ r_{oz_2} = 0;\ P_x = P_y = P_z = 0$$

$$r_{ox_3} = -a,\ r_{oy_3} = 0,\ r_{oz_3} = 0;\ P_x = P_y = P_z = 0$$

$$r_{ox_4} = 0,\ r_{oy_4} = -a,\ r_{oz_4} = 0;\ P_x = P_y = P_z = 0$$

the total torque components become

$$G_{bx}/m = P[\dot{\omega}_z a - \dot{\omega}_x \dot{P} - 2\omega_x \dot{P} + \omega_y(\omega_x a + \omega_z P)]$$

$$- 2a[\dot{\omega}_x + \omega_y \omega_z a] \tag{24}$$

$$G_{by}/m = a[\ddot{P} - \omega_y a + \omega_z(\omega_x a + \omega_z P) - P(\omega_x^2 + \omega_y^2 + \omega_z^2)]$$

$$- [P\dot{\omega}_y P + \omega_x(\omega_x a + \omega_z P) - a(\omega_x^2 + \omega_y^2 + \omega_z^2)$$

$$+ 2\omega_y \dot{P}] - a[\dot{\omega}_y a - \omega_z \omega_x a] \tag{25}$$

$$G_{bz}/m = -a[\dot{\omega}_z a - \dot{\omega}_x P + \omega_y(\omega_x a + \omega_z P) - 2\omega_x \dot{P}]$$

$$+ 2a[-\dot{\omega}_z a + \omega_x \omega_y a] + a[-\dot{\omega}_z a - \omega_y \omega_x a] \tag{26}$$

The last term in brackets in each of the equations (24) - (26) are torques due to balancing masses which have been added to get the total torque due to both damper and three balancing masses.

We separate equations (24) - (26) into two terms such that $G = G_o + G^*$ where G_o are terms due to rigid body which are directly addable to I_x, I_y and I_z respectively, and G^* is then torque due to vibrating system. Then

$$G_{bx_o}/m = -2a^2[\dot{\omega}_x + \omega_y \omega_z]$$

$$G_{by_o}/m = -2a^2[\dot{\omega}_y - \omega_z \omega_x]$$

$$G_{bz_o}/m = -4a^2[\dot{\omega}_z]$$

and

$$G^*_{bx}/m = P[\dot{\omega}_z a - \dot{\omega}_x P - 2\omega_x \dot{P} + \omega_y(\omega_x a + \omega_z P)] \quad (27)$$

$$G^*_{by}/m = a[\ddot{P} - P(\omega_x^{\ 2} + \omega_y^{\ 2})] - P[\dot{\omega}_y P + 2\omega_y \dot{P} + \omega_x\omega_z P - a(\omega_y^{\ 2} + \omega_z^{\ 2})] \quad (28)$$

$$G^*_{bz}/m = -a[-\dot{\omega}_x P + \omega_y\omega_z P - 2\omega_x \dot{P}] \quad (29)$$

Hence, $G_a = +E_{as}G_b$ where E_{as} is as defined in (2), and the components of torque are

$$G_{ax} = G_{by}^{\ *} c\alpha - G_{by}^{\ *} s\alpha \quad (30)$$

$$G_{ay} = G_{by}^{\ *} c\alpha + G_{bx}^{\ *} s\alpha \quad (31)$$

$$G_{az} = G_{bz}^{\ *} \quad (32)$$

We shall now derive bearing torques. The torque on the AMCD rim about the axis of rotation resulting from magnetic bearing forces F_1, F_2, and F_3 can be expressed in terms of axial magnetic bearing gains k_ϕ and $k_{\dot{\phi}}$. The bearing forces are defined by the equations (see Anderson and Groom [1]).

$$F_1 = r(k_\phi \phi + k_{\dot{\phi}} \dot{\phi})\sin\alpha + k_{\dot{\phi}} \phi r \dot{\alpha} \cos\alpha$$

$$F_2 = -r(k_\phi \phi + k_{\dot{\phi}} \dot{\phi})\sin(60^\circ + \alpha) - k_{\dot{\phi}} \phi r \dot{\alpha} \cos(60^\circ + \alpha)$$

$$F_3 = r(k_\phi \phi + k_{\dot{\phi}} \dot{\phi})\sin(60^\circ - \alpha) - k_{\dot{\phi}} \phi r \dot{\alpha} \cos(60^\circ - \alpha)$$

The torque is the resultant of the individual torques produced by the bearing forces:

$$T = (-F_1 \sin\theta + F_2 \sin(60^\circ + \theta) - F_2 \sin(60^\circ - \theta))r$$

Substituting and simplifying we have $T = -k_{\dot{\phi}}\dot{\phi} - k_{\dot{\phi}}\dot{\phi}$ where $k_{\phi} = 1.5\ r^2 k_{\phi}$ and $k_{\dot{\phi}} = 1.5\ r^2 k_{\dot{\phi}}$.

Thus, the net bearing torque is not dependent on the position of the magnetic bearing segments relative to the rim (the torque is independent of α) and, therefore, the three linear bearings may be treated as two rotational bearings.

The bearing torques can now be written as

$$T_{sx} = k_{\phi}(\phi_a - \phi_s) + k_{\dot{\phi}}(\dot{\phi}_a - \dot{\phi}_s) + k_{\dot{\phi}}\dot{\beta}(\theta_a - \theta_s)$$

$$T_{sy} = k_{\phi}(\theta_a - \theta_s) + k_{\dot{\phi}}(\dot{\theta}_a - \dot{\theta}_s) - k_{\dot{\phi}}\dot{\beta}(\phi_a - \phi_s)$$

The spin motor torque T_{sz} can be used to control either spacecraft z-axis attitude or attitude rate, as well as to counteract rim drag torque (hypteresis and eddy current losses). In this derivation, the component of motor torque, which is greater than the drag torque, is of interest and $T_{sz} = T_c - T_d$ where subscripts c and d refer to control and drag. Noting that

$$T_{az} = -E_{as}\ T_{sz} \tag{33}$$

and assuming small angular motion for the case where $T_c = T_d$ and $G = 0$, the equation (33), with $E_{as} = I$ (unity), becomes

$$T_{ax} = -T_{sx} = -k_{\phi}(\phi_a - \phi_s) - k_{\dot{\phi}}(\dot{\phi}_a - \dot{\phi}_s) - k_{\dot{\phi}}\dot{\beta}(\theta_a - \theta_s) \tag{34}$$

$$T_{ay} = -T_{sy} = -k_{\phi}(\theta_a - \theta_s) - k_{\dot{\phi}}(\dot{\theta}_a - \dot{\theta}_s) + k_{\dot{\phi}}\dot{\beta}(\phi_a - \phi_s) \tag{35}$$

$$T_{az} = -T_{sz} = 0 \text{ for no friction} \tag{36}$$

STATE EQUATION AND STABILITY CRITERION

Eliminating torques T_{ax}, T_{ay}, T_{az} and T_{sx}, T_{sy}, T_{sz} in equations (14) - (21), and using equations (34) - (36) yields the following set of AMCD spacecraft equations:

$$I_a\ddot{\phi}_a = -k_\phi(\phi_a - \phi_s) - k_{\dot{\phi}}(\dot{\phi}_a - \dot{\phi}_s) - H_a\dot{\theta}_a$$
$$- k_{\dot{\phi}}\dot{\beta}(\theta_a - \theta_s) + G_{ax} \tag{37}$$

$$I_a\ddot{\theta}_a = -k_\phi(\theta_a - \theta_s) - k_{\dot{\phi}}(\dot{\theta}_a - \dot{\theta}_s) - H_a\dot{\phi}_a$$
$$+ K_{\dot{\phi}}\dot{\beta}(\phi_a - \phi_s) + G_{ay} \tag{38}$$

$$I_s\ddot{\phi}_s = k_\phi(\phi_a - \phi_s) + k_{\dot{\phi}}(\dot{\phi}_a - \dot{\phi}_s) - H_s\dot{\theta}_s$$
$$+ k_{\dot{\phi}}\dot{\beta}(\theta_a - \theta_s) + G_{sx} \tag{39}$$

$$I_s\ddot{\theta}_s = k_\phi(\theta_a - \theta_s) + k_{\dot{\phi}}(\dot{\theta}_a - \dot{\theta}_s) + H_s\dot{\phi}_s$$
$$- k_{\dot{\phi}}\dot{\beta}(\phi_a - \phi_s) + G_{sy} \tag{40}$$

$$\dot{H}_a = G_{az} = I_{az}\ddot{\alpha} \tag{41}$$

$$\dot{H}_s = G_{sz} = I_{sz}\ddot{\beta} \tag{42}$$

Equations of motion for dampers are given by

$$c\dot{P} + kP = F_{az} \tag{43}$$

$$c'\dot{P}' + k'P' = F'_{az} \tag{44}$$

where F_{az} is force along z-axis and is obtained from equations (22) and (23) as

$$F_{az} = -m\{\ddot{P} - \dot{\omega}_y a + \omega_z\omega_x a - P(\omega_x{}^2 + \omega_y{}^2)\} \tag{45}$$

and similarly

$$F_{sz}' = -m'\{\ddot{P}' - \omega_y'a' + \omega_z'\omega_x'a' - P'(\omega_x'^2 + \omega_y'^2)\} \tag{46}$$

where G_{zx}, G_{ay}, and G_{az} given by equations (30) - (32). With the subscript changed, and omitting the algebra, we can write $G_{sx} = G_{bx}'^{*}c\beta - G_{by}'^{*}s\beta$, $G_{sy} = G_{by}'c\beta + G_{bx}'^{*}s\beta$, and $G_{sz} = G_{bz}'^{*}$ where $G_{bx}'^{*}$, $G_{by}'^{*}$ and $G_{bz}'^{*}$ are derived similar to equations (27) - (29). Equations (37) - (46) are linearized and are written in the form

$$I(t)\ddot{x}(t) + B(t)\dot{x}(t) + C(t)x(t) = 0 \tag{47}$$

where the vector $x = (\phi_a, \theta_a, \phi_s, P, P')$, and $I(t)$, $B(t)$ and $C(t)$ are matrices

$$I(t) = \begin{bmatrix} I_a & 0 & 0 & 0 & a_{15} & 0 \\ 0 & I_a & 0 & 0 & a_{25} & 0 \\ 0 & 0 & I_s & 0 & 0 & a_{36} \\ 0 & 0 & 0 & I_s & 0 & a_{46} \\ a_{51} & a_{52} & 0 & 0 & m & 0 \\ 0 & 0 & a_{63} & a_{64} & 0 & m' \end{bmatrix}$$

where

$$a_{15} = a_{51} = mas\alpha, \ a_{25} = a_{52} = -mac\alpha$$

$$a_{36} = a_{63} = m'a's\beta, \ a_{46} = a_{64} = -m'a'c\beta$$

$$B(t) = \begin{bmatrix} k_{\dot{\phi}} & I_{az}\omega_{\alpha o} & -k_{\dot{\phi}} & 0 & 0 & 0 \\ -I_{az}\omega_{\alpha o} & k_{\dot{\phi}} & 0 & -k_{\dot{\phi}} & 0 & 0 \\ -k_{\dot{\phi}} & 0 & k_{\dot{\phi}} & I_{sz}\omega\beta_{o} & 0 & 0 \\ 0 & -k_{\dot{\phi}} & -I_{sz}\omega_{\beta o} & k_{\dot{\phi}} & 0 & 0 \\ 2a\omega_{\alpha o}mc\alpha & 2a\omega_{\alpha o}ms\alpha & 0 & 0 & c & 0 \\ 0 & 0 & 2a'\omega_{\beta o}m'c\beta & 2a'\omega_{\beta o}m's\beta & 0 & c' \end{bmatrix}$$

$$C(t) = \begin{bmatrix} k_\phi & k_{\dot\phi}\omega_{\beta o} & -k_\phi & k_{\dot\phi}\omega_{\beta o} & ma\omega_{\alpha o}^2 s\alpha & 0 \\ -k_{\dot\phi}\omega_{\beta o} & k_\phi & k_{\dot\phi}\omega_{\beta o} & -k_\phi & -ma\omega_{\alpha o}^2 c\alpha & 0 \\ -k_\phi & -k_{\dot\phi}\omega_{\beta o} & k_\phi & k_{\dot\phi}\omega_{\beta o} & 0 & m'a'\omega_{\beta o}^2 s\beta \\ k_{\dot\phi}\omega_{\beta o} & -k_\phi & k_{\dot\phi}\omega_{\beta o} & k_\phi & 0 & -m'a'\omega_{\beta o}^2 c\beta \\ 0 & 0 & 0 & 0 & k & 0 \\ 0 & 0 & 0 & 0 & 0 & k' \end{bmatrix}$$

Case Without Dampers

The system without dampers is obtained by deleting the last two equations (43) - (44) in the state equations, since they represent the equations of motions for the dampers. We then write the equations in the form

$$I\ddot{x}(t) + (B_1 + B_2)\dot{x}(t) + (C_1 + C_2)x(t) = 0 \tag{48}$$

where

$$B_1 = \begin{bmatrix} k_\phi & 0 & -k_{\dot\phi} & 0 \\ 0 & k_{\dot\phi} & 0 & -k_{\dot\phi} \\ -k_{\dot\phi} & 0 & k_{\dot\phi} & I_{sz}\omega_{\beta o} \\ 0 & -k_{\dot\phi} & -I_{sz}\omega_{\beta o} & k_{\dot\phi} \end{bmatrix}$$

$$B_2 = \begin{bmatrix} 0 & I_{ax}\omega_{\alpha o} & 0 & 0 \\ -I_{az}w_{\alpha o} & 0 & 0 & 0 \\ 0 & 0 & 0 & 0 \\ 0 & 0 & 0 & 0 \end{bmatrix}$$

$$C_1 = \begin{bmatrix} k_\phi & 0 & -k_\phi & -k_{\dot\phi}\omega_{\beta o} \\ 0 & k_\phi & k_{\dot\phi}\omega_{\beta o} & -k_\phi \\ -k_\phi & -k_\phi\omega_{\beta o} & k_\phi & k_{\dot\phi}\omega_{\beta o} \\ k_{\dot\phi}\omega_{\beta o} & -k_\phi & -k_{\dot\phi}\omega_{\beta o} & k_\phi \end{bmatrix}$$

and

$$C_2 = \nu B_2 \text{ where } \nu = k_{\dot\phi}\omega_{\beta o}/I_{az}\omega_{\alpha o}$$

The results with no dampers are summarized into the following theorem:

THEOREM 1. If $I > 0$, $C_1 \geq 0$, $B_1 - \nu I \geq 0$, then the system (48) is asymptotically stable provided

$$C_1 + \nu B_1 - (1+\lambda)\nu^2 I/\lambda > 0$$

for some $\lambda > 0$, and $w_{\beta o} \equiv 0$.

Proof. Consider a Liapunov function

$$V = \dot{x}^T I \dot{x} + (1 + \lambda)x^T[C_1 + \nu B_1 - (1+\lambda)\nu^2 I/\lambda]x$$
$$+ \lambda[(1+\lambda)\nu x/\lambda + \dot{x}]^T I[(1+\lambda)\nu x/\lambda + \dot{x}] \qquad (49)$$

whose time derivative along the trajectories of (48) yields:

$$-\dot{V} = 2\nu(1 + \lambda)x^T C_1 x + 2(1 + \lambda)\dot{x}^T[B_1 - \nu I]\dot{x}$$

By assumption, $V > 0$ and $\dot{V} \leq 0$, which implies that the system is asymptotically stable by LaSalle's Theorem [12].

Remark 1. If $I > 0$, $B_1 + B_2 = B > 0$, and $C_1 + C_2 = C \geq 0$ then the system (48) is asymptotically stable. The result follows by choosing $\nu = 0$ and Liapunov-function in (49).

We next derive the necessary condition for the system (47) to be asymptotically stable. We first state a well-known lemma.

LEMMA 2. Halanay [8]. If the characteristic roots of M in the equation $\dot{x} = Mx$ are stable then the equation

$$M^T S + SM = -C \quad (C > 0)$$

is unique and is given by the formula

$$S = \int_0^\infty e^{M^T t} C e^{Mt} dt$$

Conversely, if $C > 0$ and $S > 0$ then M is stable.

THEOREM 3. (Necessary Condition). If (48) is asymptotically stable then $B > 0$ and $C \geq 0$.

Proof. Consider a function

$$V = x^T Sx, \quad S = \begin{bmatrix} S_{11} & S_{12} \\ S_{12}^T & S_{22} \end{bmatrix} > 0$$

such that

$$\dot{V} = -x^T Qx; \quad Q = \begin{bmatrix} Q_{11} & 0 \\ 0 & Q_{22} \end{bmatrix}$$

A simple computation of $\dot{V}$ gives $C^T S_{12} + S_{12} C = Q_{11}$, $-S_{12} - S_{12}^T + B^T S_{22} + S_{22} B = Q_{22}$, $S_{11} - S_{12} B - C^T S_{22} = 0$, and $S_{11} - B^T S_{12}^T - S_{22} C = 0$. If in $-C$ the real parts of characteristic roots are positive, then S_{12} can be uniquely determined from Lemma 2. Therefore, from the second set of equations $-B^T S_{22} - S_{22} B = -Q_{22} - S_{12} - S_{12}^T < 0$. Therefore, $-B$ must be stable if $S_{22} > 0$.

Case With Dampers

The passive AMCD-spacecraft equations of motion with one damper and three symmetrical masses on AMCD and spacecraft are represented by the state equations (47). To study the stability properties of the system (47), we represent it in the form

$$\dot{x} = M(t)x \tag{50}$$

where

$$M = \begin{bmatrix} 0 & I \\ q_1 & p_1 \end{bmatrix}, \; p_1 = -I^{-1}(t)B(t), q_1 = -I^{-1}(t)c(t)$$

We rewrite equation (50) as

$$\dot{x} = [A + B(t)]x \tag{51}$$

such that the matrices A and B(t) satisfy the conditions of Theorem 5, given subsequently.

The following lemma gives a constant positive definite matrix S for the system (50). The result for constant matrix B is given in [26].

LEMMA 4. There exists a constant matrix $\tilde{S}$ satisfying

$$SB(t) + B^T(t)S = 0$$

if and only if

$$\tilde{S} = \lim_{T\to\infty} \frac{1}{2T} \int_{-T}^{T} X^T(s,t)PX(s,t)ds$$

exists and is finite, is independent of t, and is nonsingular for some positive matrix P.

Proof. Let $x(t,t_0)$ be the fundamental matrix of B(t) such that $X(t_0,t_0) = I$. If s exists and is constant, then $X^T(s,t)sX(s,t) = s$ and $s = \tilde{s}$. If $\tilde{s}$ exists and is independent of t, then by pre-multiplying with $X(t,t_0)$ and post-multiplying by $X(t,t_0)$, we get

$$X^T(t,t_0)\tilde{s}X(t,t_0) = \lim_{T\to\infty} \frac{1}{2T}\int_{-T}^{T} X^T(s,t_0)PX(s,t_0)ds$$

$\tilde{s}$ is clearly symmetric and non-negative, but it is assumed to be nonsingular, so it must be positive-definite.

THEOREM 5. Suppose that $A \leq 0$ and the matrix $B(t)$ satisfies the conditions of Lemma 4; then the system (51) is stable.

Proof. Consider the V-function $V = x^T sx$ whose time derivative along the trajectories of (51) is $\dot{V} = x^T[B^T s + sB]x + x^T[A^T s + sA]x = x^T[A^T s + sA]x \leq 0$. Stability of the system follows from LaSalle's Theorem.

Discussion. The stability criterion obtained was simulated with the parameters of a prototype AMCD-spacecraft model under investigation at Langley Research Center, NASA. The system was found to be asymptotically stable. The parameters used were the same as those given by Anderson and Groom [1]: I_a = 680 kg-m^2, I_{az} = 1360 kg-m, I_s = 680 kg-m^2, I_{sz} = 453.3 kg-m^2, m = .15 kg, m' = .15 kg, a = 0.76 m, a' = 0.76 m, $\omega_{\alpha o}$ = 401.3 rad/sec, $\omega_{\beta o}$ = 0.1 rad/sec, k = 1.2, k' = 4.1412, c = .16, c' = .0688, k_ϕ = 1020, $k_{\dot{\phi}}$ = 2856.0. Simulation was repeated with k_ϕ = 1360 and $k_{\dot{\phi}}$ = 3808. The results for the case with no dampers was simulated.

The results presented in this paper are concerned with a reasonably realistic model of an important type of attitude control system. They answer key questions concerning the stability of the system, and provide an analytical basis for using a computer for further studies.

On the other hand, although we have proved the basic stability properties of the system, in this paper we have not considered the next natural problem, that of determining the extent to which the system performance can be improved as a result of the presence of the non-linearities. There are several other important practical problems that are not considered here such as: (i) the problem of predicting a control law which will stabilize the system; (ii) other design considerations; (iii) the problem of comparing the performance with alternative systems. Considerable amount of work is needed before NASA will use AMCD for spacecraft attitude control system.

ACKNOWLEDGEMENT

Part of this work was done at Langley Research Center. The author is indebted to W. W. Anderson and his associates of the Langley Research Center for their many helpful suggestions.

REFERENCES

1. W. W. Anderson and N. J. Groom, The annular momentum control device and applications, NASA TN D-7866, March 1975.
2. P. M. Bainum, P. G. Fuechsel, and D. L. Mackinson, On the motion and stability of a dual-spin satellite with nutation damping, Applied Physics Laboratory, John Hopkins Univ., TG-1072, 1969.
3. L. Cesari, Asymptotic Behavior and Stability Problems in Ordinary Differential Equations, Springer-Verlag, New York, 1963.
4. B. T. Fang, Energy considerations for attitude stability of dual-spin spacecraft, J. Spacecraft and Rockets 5 (1968), 1241-1243.
5. T. W. Flatley, Equilibrium states for a class of dual-spin spacecraft, NASA TR R-362, 1971.
6. H. Goldstein, Classical Mechanics, Addison-Wesley, Reading, 1957.
7. N. J. Groom, Simplified analytical model of an annular momentum control device two-axis passive control system, NASA, LWP-1130, September, 1973.
8. A. Halanay, Differential Equations: Stability, Oscillations, Time Lags, Academic Press, New York, 1966.
9. A. J. Iorillo, Nutation damping dynamics of axis-symmetric rotor stabilized satellites, ASME Winter Meeting, Chicago, Illinois, 1965.
10. T. P. Kane and P. M. Barba, Effects of energy dissipation of a spinning satellite, AIAA J. 4 (1966), 1391-1393.
11. Kurzhals, P. R., Spin dynamics of space stations under transient and steady-state excitations and stabilizing responses, M. S. Thesis, Virginia Polytechnic Inst., Blacksburg, 1962.

12. J. P. LaSalle, An Invariance Principle in the Theory of Stability, Division of Appl. Math., Brown University, TR 66-1, Providence, 1966.

13. P. W. Likins, Effects of energy dissipation on the free body motions of spacecraft, Jet Propulsion Laboratory Technical Report No. 32-860, 1966.

14. P. W. Likins, Attitude stability of dual-spin system, Hughes Aircraft, Space System Division, SSD 63077R, 1966.

15. P. W. Likins, Stability theory and results, presented at the symposium on Attitude Stabilization and Control of Dual Spin Spacecraft, El Segundo, California, August 1-2, 1967.

16. P. W. Likins, Attitude stability criteria for dual-spin spacecraft, J. of Spacecraft and Rockets 4 (1967), 1638-1643.

17. P. W. Likins, Gan Tai Tsing, and D. L. Mingori, Stable limit cycles due to nonlinear damping in dual-spin spacecraft, AIAA Paper No. 70-1044, 1970.

18. R. J. McElvain and W. W. Porter, Design consideration for spin axis control of dual-spin spacecraft, presented at the Symposium on Attitude Stabilization and Control of Dual-Spin Spacecraft, El Segundo, California, August 1-2, 1967.

19. L. Meirovitch, Attitude stability of an elastic body of revolution in space, J. Astronaut Sc. 8 (1961), 110-113.

20. D. L. Mingori, The determination of floquet analysis of the effects of energy dissipation on the attitude stability of dual-spin satellites, Proc. Symposium Attitude Stabilization and Control of Dual-Spin Spacecraft, Air Force Report No. SMASO-TR-68-191, 1967.

21. D. L. Mingori, Effects of energy dissipation on the attitude stability of dual-spin satellites, AIAA 71 (1969), 20-27.

22. H. Perkel, Stability, A Three Axis Attitude Control System Utilizing a Single Reaction Wheel, AIAA Paper No. 66-307, 1966.

23. R. Pringle, Jr., Stability of damped mechanical systems, AIAA 2 (1965), 363-364.

24. C. S. Rall, Nutational instabilities in dual-spin satellites resulting from vehicle flexibility, Aerospace Corporation Report TOP-0158 (3133-01)-01, 1960.

25. A. K. Sen, Stability of a dual-spin satellite with a four-mass nutation damper, AIAA Journal 8 (1970), 822-823.

26. A. S. C. Sinha, Global asymptotic stability of certain nonlinear feedback systems, Proc. IEEE 61 (1973), 1506-1507.

27. W. T. Thompson and G. S. Reiter, Motion of an asymmetric spinning body with internal dissipation, AIAA Journal 6 (1963), 1429-1430.

Chapter 18

THE IMPLICIT FUNCTION THEOREM AND A RANDOM INTEGRAL EQUATION

JAMES R. WARD*

Department of Mathematics
University of Oklahoma
Norman, Oklahoma

INTRODUCTION

One of the many approaches to random integral and differential equations is to extend the methods of functional analysis to random problems ([1,8]). Here we apply one of the most powerful methods of nonlinear functional analysis to a random Volterra integral equation of the form

$$x(t;\omega) = y(t;\omega) + \int_0^t a(t,\tau;\omega)g(\tau,x(\tau;\omega))d\tau \qquad (1)$$

The method we speak of is the application of the Hildebrandt-Graves implicit function theorem (see [2] or [11]) for equations in a Banach space. We use the implicit function theorem to obtain existence (for $t \in R^+ = [0,\infty)$) and stability results for (1). This method has been widely used to study the existence and qualitative behavior of solutions to deterministic equations; see, e.g., [3, 9, or 10].

---

*Present address: Department of Mathematics, Pan American University, Edinburg, Texas.

PRELIMINARIES

Let R denote the set of real numbers, $R^+ = [0,\infty)$, and R^n the set of n-dimensional real column vectors. For $x \in R^n$ we use $|x|$ to denote the absolute value of x if n = 1, and to denote any convenient vector norm if n > 1. If A is any n x n matrix let $|A| = \sup_{|x|=1} |Ax|$, $x \in R^n$.

We will assume that $\omega \in \Omega$ is an element in the supporting set of a complete probability measure space (Ω, A, μ). For $1 \leq p < \infty$ and $n \geq 1$ we use $L_p^n(\Omega)$ to denote the space of all A-measurable R^n-valued functions $x(\omega)$ defined on Ω such that

$$||x(\omega)||_p = \{\int_\Omega |x(\omega)|^p d\mu\}^{1/p} < \infty$$

By $L_\infty^n(\Omega)$ we mean the space of all A-measurable R^n-valued functions $x(\omega)$ such that

$$||x(\omega)||_\infty = \mu\text{-ess} \sup_\Omega |x(\omega)| < \infty$$

$C(R^+, L_p^n(\Omega))$ will denote the set of all continuous functions $x(t;\omega)$ on R^+ with values in $L_p^n(\Omega)$. $C_b(R^+, L_p^n(\Omega))$ will denote the Banach space of all continuous functions $x(t;\omega)$ on R^+ into $L_p^n(\Omega)$ which are bounded in $L_p^n(\Omega)$ with norm

$$||x||_b = \sup_{t \geq 0} ||x(t;\omega)||_p \tag{2}$$

STATEMENT OF RESULTS

We use the following result which is an extension of the classical theory of Volterra equations. Let $\Delta = \{(t,\tau) : 0 \leq \tau \leq t < \infty\}$.

PROPOSITION 1. Let X be a Banach space with norm $||.||$, L(X) the space of bounded linear operators on X with the uniform operator norm. If $y \in C(R^+, X)$ and $a \in C(\Delta, L(X))$, with $a(t,\tau) = 0$ if $\tau > t$, then there is a unique solution $x \in C(R^+, X)$ to the linear equation

$$x(t) = y(t) + \int_0^t a(t,\tau)x(\tau)d \tag{3}$$

Moreover, x(t) is given by

$$x(t) = y(t) - \int_0^t r(t,\tau)y(\tau)d$$

where $r \in C(\Delta, L(X))$ is the unique continuous solution to the operator equation

$$r(t,\tau) = -a(t,\tau) + \int_\tau^t a(t,u)r(u,\tau)du \tag{4}$$

We will omit the proof of Proposition 1, as the method is essentially the same as that for the classical case of X = R, see [7], with norms replacing absolute values. The operator valued function $r(t,\tau)$ is called the resolvent kernel for $a(t,\tau)$ or the kernel reciprocal to $r(t,\tau)$.

COROLLARY 2. Suppose $a = a(t,r;\omega) \in C(\Delta, M_\infty^n(\Omega))$ where $M_\infty^n(\Omega)$ is the space of $n \times n$ random matrices $A(\omega)$ all of whose components are in $L_\infty^n(\Omega)$. Then if $y(t;\omega) \in C(R^+, L_p^n(\Omega))$, $1 \le p \le \infty$, there exists a unique $x(t;\omega) \in C(R^+, L_p^n(\Omega))$ such that

$$x(t;\omega) = y(t;\omega) + \int_0^t a(t,\tau;\omega)x(\tau;\omega)d\tau \tag{5}$$

and

$$x(t,\omega) = y(t;\omega) - \int_0^t r(t,\tau;\omega)y(t;\omega)d\tau \tag{6}$$

where $r(t,\tau;\omega)$ is given by (4).

The proof follows directly from Proposition 1 and the observation that $a(t,\tau;\omega)$ is a bounded linear operator on $L_p^n(\Omega)$ for each $(t,\tau) \in \Delta$, and the map $(t,\tau) \to a(t,\tau;\omega)$ from Δ into $L(L_p^n(\Omega))$ is continuous.

If there is a number $c > 0$ such that

$$\sup_{t\geq 0}\int_0^t ||a(t,\tau;\omega)||_\infty d\tau + \sup_{t\geq 0}\int_0^t ||r(t,\tau;\omega)||_\infty d\tau \leq c \qquad (7)$$

then for each $y(t;\omega) \varepsilon C_b(R^+,L_p{}^n(\Omega))$ the solution $x(t;\omega)$ of equation (5) is in the same space, and the mapping $y \to x$ is an invertible bounded linear operator on $C_b(R^+,L_p{}^n(\Omega))$.

We now consider the nonlinear equation (1). Let $g : R^+ \times R^n \to R^n$ be continuous on $R^+ \times R^n$ and continuously differentiable with respect to $x \varepsilon R^n$ for all $(t,x) \varepsilon R^+ \times R^n$. We write the Jacobian matrix $\frac{\partial}{\partial x} g(t,x) = g_x'(t,x)$. Also assume $g_x'(t,x)$ is continuous in x uniformly with respect to $t \varepsilon R^+$ for x in compact subsets of R^n, and that $g(t,0) = 0$.

If $g(t,x)$ satisfies these assumptions then the mappings G and G_x on $C_b(R^+,L_\infty{}^n(\Omega))$ given by $G(u)(t,\omega) = g(t,u(t,\omega))$ and $G_x(u)(t,\omega) = g_x'(t,u(t;\omega))$ are each continuous mappings on $R^+ \times L_\infty{}^n(\Omega)$ into $L_\infty{}^n(\Omega)$. These facts are used in the proof of the following theorem which is the main result of this chapter.

THEOREM 3. Let $g(t,x)$ satisfy the above assumptions, and let $a(t,\tau;\omega) \varepsilon C(\Delta,M_\infty{}^n(\Omega))$. Let $r_o(t,\tau;\omega)$ denote the kernel reciprocal to $a(t,\tau;\omega)g_x'(t,0)$. If

$$\sup_{t\geq 0}\int_0^t ||a(t,\tau;\omega)g_x'(t,0)||_\infty d\tau$$

$$+ \sup_{t\geq 0}\int_0^t ||r_o(t,\tau;\omega)||_\infty d\tau = c < \infty \qquad (8)$$

then there are numbers $\alpha > 0$ and $\beta > 0$ and an (solution) operator $T : S(\alpha) = \{y \varepsilon C_b(R^+,L_\infty{}^n(\Omega)) : ||y||_b < \alpha\} \to S(\beta)$ such that

(i) $T(y) = x$ solves equation (1) for each $y \varepsilon S(\alpha)$, and moreover, there is no other solution in the ball $S(\beta)$;

(ii) T is continuously Frechét differentiable on $S(\alpha)$ and $T'(y)h = z$ is the solution to the linear equation

$$z(t;\omega) = h(t;\omega) + \int_0^t a(t,\tau;\omega)g_x'(\tau,x(\tau;\omega))z(\tau;\omega)d\tau$$

We omit the proof of Theorem 3, which follows the following scheme: Define the operator F on $C(R^+,L_\infty{}^n(\Omega))$ by

$$F(u)(t;\omega) = u(t;\omega) - \int_0^t a(t,\tau;\omega)g(\tau,u(\tau;\omega))d\tau$$

One then shows that F maps $C_b(R^+,L_\infty{}^n(\Omega))$ into itself and is continuously Fréchet differentiable on $C_b(R^+,L_\infty{}^n(\Omega))$, with $F'(u)h$, for $h \in C_b(R^+,L_\infty{}^n(\Omega))$ given by

$$F'(u)h(t;\omega) = h(t;\omega) - \int_0^t a(t,\tau;\omega)g_x'(\tau,u(\tau;\omega))h(\tau;\omega)d\tau$$

Then one uses (8) and the remarks following (7) to show that $F'(0)$ is invertible in $L(L_\infty{}^n(\Omega))$. One then has (1) F is continuously Fréchet differentiable, (2) $F'(0)$ is invertible, and (3) $F(0) = 0$. These are the hypotheses of the implicit function theorem (see [2] or [11]); the conclusions of Theorem 3 are the conclusions of the implicit function theorem interpreted for equation (1).

COROLLARY 4. If the hypotheses of Theorem 3 are satisfied, then equation (1) is stable in the sense that if $\varepsilon > 0$ is given there is a number $\delta > 0$ such that if $y \in C_b(R^+,L_\infty{}^n(\Omega))$ with $||y||_b < \delta$ then there is a solution $x \in C_b(R^+,L_\infty{}^n(\Omega))$ with $||x||_b < \varepsilon$.

Proof. The operator T is continuously Frechét differentiable on $S(\alpha)$ and is therefore continuous there; in particular, it is continuous at the origin.

It would be interesting to see if Theorem 3 can be extended to the case of $y \in C_b(R^+,L_p{}^n(\Omega))$, $1 \leq p < \infty$. That such an extension is not straightforward can perhaps be seen by the observation that if $g : R^+ \times R^n \to R^n$ and for each $t \geq 0$ the operator G(t) is given by $G(t)u(\omega) = g(t,u(\omega))$, then G(t) may map all of $L_2^n(\Omega)$ into itself, but such an operator can never be Frechet differentiable at any point of $L_2^n(\Omega)$ (cf. [9]).

REFERENCES

1. A. T. Bharucha-Reid, Random integral equations, Academic, New York, 1972.
2. J. Dieudonné, Foundations of modern analysis, Academic, New York, 1969.
3. A. G. Kartsatos, The Hildebrandt-Graves theorem and the existence of solutions to boundary value problems on infinite intervals, Math. Nachr. 67 (1975), 91-100.
4. M. A. Krasnosel'skii, Topological methods in the theory of nonlinear integral equations, Macmillan, New York, 1964.
5. J. S. Milton and C. P. Tsokos, On a class of nonlinear stochastic integral equations, Math. Nachr. 60 (1974), 61-78.
6. J. L. Strand, Random ordinary differential equations, J. of Differential Equations 7 (1970), 538-553.
7. F. G. Tricomi, Integral equations, Interscience, New York, 1956.
8. C. P. Tsokos and W. J. Padgett, Random integral equations with applications to stochastic systems, Springer-Verlag, New York, 1971.
9. M. M. Vainberg, Variational methods for the study of nonlinear operators, Holden-Day, San Francisco, 1964.
10. J. R. Ward, The existence of solutions, stability, and linearization of Volterra systems, to appear.
11. T. H. Hildebrandt and L. M. Graves, Implicit functions and their differentials in general analysis, Trans. Amer. Math. Soc. 29 (1927), 127-153.

Chapter 19

ASYMPTOTIC STABILITY IN THE α-NORM FOR AN ABSTRACT NONLINEAR VOLTERRA INTEGRAL EQUATION

G. F. Webb
Department of Mathematics
Vanderbilt University
Nashville, Tennessee

INTRODUCTION

Let X be a Banach space with norm $\|\ \|$. Our objective is to discuss the asymptotic behavior of solutions to the abstract semi-linear differential equation in X

$$du(t)/dt = -Au(t) + B(t,u(t)), \quad t > t_0 \qquad (1)$$
$$u(t_0) = x \in X$$

where -A is the infintesimal generator of a strongly continuous holomorphic semigroup of linear operators in X and $B(t,\cdot)$ is a nonlinear operator defined on the domain of a fractional power of A to X. The equation (1) has been studied extensively in this framework and it is known that local solutions exist under reasonable general continuity assumptions on the non-linear term B (see [1, 2, and 4]). In our study we shall assume the existence of a global solution of (1) satisfying an a priori bound. Roughly speaking, our results say that if the linearized equation $du(t)/dt = -Au(t)$ is asymptotically stable, and if the nonlinear perturbation B is sufficiently small, then the semi-linear equation (1) is asymptotically stable.

We make the following assymptions on A and B:

H_1. $-A$ is the infinitesimal generator of a strongly continuous holomorphic semigroup of linear operators $T(t)$, $t \geq 0$ in X;

H_2. there exist constants $M \geq 1$ and $\omega < 0$ such that $\|T(t)x\| \leq Me^{\omega t}\|x\|$ for $t \geq 0$, $x \in X$;

H_3. there exists $\alpha \in (0,1)$ such that A^{α} is 1-1 and onto, so that $A^{-\alpha}$ is bounded and everywhere defined and $D(A^{\alpha}) \overset{def}{=} X_{\alpha}$ is a Banach space with norm $\|x\|_{\alpha} = \|A^{\alpha}x\|$, $x \in X_{\alpha}$;

H_4. $B : [t_0,\infty) \times X_{\alpha} \to X$.

Under the assumptions above it is advantageous to study the integrated version of equation (1) given by the singular nonlinear Volterra integral equation

$$u(t) = T(t-t_0)x + \int_{t_0}^{t} T(t-s)B(s,u(s))ds, \quad t \geq t_0 \tag{2}$$

If $B(t,u(t))$ is Hölder continuous in t for $t \geq t_0$, then $u(t)$ as given in (2) is continuous for $t \geq t_0$, continuously differentiable for $t > t_0$, and satisfies (1) (see [3; Theorem 1.27]). In our treatment we shall suppose that $x \in D(A^{\alpha})$ and study the asymptotic properties of the solutions of equation (2) in the α-norm, that is, in the space X_{α}.

ASYMPTOTIC BEHAVIOR OF THE SOLUTIONS

<u>THEOREM.</u> Suppose H_1 - H_4 hold, $x \in D(A^{\alpha})$, and there exists a continuous function $u : [t_0,\infty) \to D(A^{\alpha})$ satisfying (2). Suppose there exists a constant $L > 0$ and a continuous function $h : [t_0,\infty) \to [0,\infty)$ such that $\|B(t,u(t))\| \leq L(\|u(t)\|_{\alpha} + h(t))$ for $t \geq t_0$. Finally, suppose that $L < (-\omega)^{1-\alpha}/C\Gamma(1-\alpha)$. Then:

(i) if $h(t)$ is bounded on $[t_0,\infty)$, then $\|u(t)\|_{\alpha}$ is bounded on $[t_0,\infty)$; (ii) if $h(t) = O(e^{\sigma t})$, where $\omega + (CL\Gamma(1-\alpha))^{1/1-\alpha} < \sigma < 0$, then $\|u(t)\|_\alpha = O(e^{\sigma t})$; (iii) if $h(t) = o(1)$, then $\|u(t)\|_\alpha = o(1)$.

Proof. We will use the gamma function formula

$$\int_0^\infty e^{-\beta t} t^{-\alpha} dt = \Gamma(1-\alpha)\beta^{\alpha-1} \text{ for } 0 < \alpha < 1,\ \beta > 0 \tag{3}$$

(see [6; p. 265]). Let γ be a real number such that $\omega + (CL\Gamma(1-\alpha))^{1/1-\alpha} < \gamma < 0$. Define $c(\gamma) = CL\Gamma(1-\alpha)(\gamma-\omega)^{\alpha-1}$ and for $t \geq t_0$ define $S_t = \sup\{e^{-\gamma s}\|u(s)\|_\alpha : t_0 \leq s \leq t\}$ and $H_t = \sup\{e^{-\gamma s}h(s) : t_0 \leq s \leq t\}$. Let $t \geq t_0$ and using (3) we obtain

$$\begin{aligned}
e^{-\gamma t}\|u(t)\|_\alpha &\leq e^{-\gamma t}\|A^\alpha T(t-t_0)x\| \\
&\quad + e^{-\gamma t}\int_{t_0}^t \|A^\alpha T(t-s)B(s,u(s))\|ds \\
&\leq e^{-\gamma t}\int_{t_0}^t Ce^{\omega(t-s)}(t-s)^{-\alpha}L(\|u(s)\|_\alpha \\
&\quad + h(s))ds + e^{-\gamma t}Me^{\omega(t-t_0)}\|x\|_\alpha \\
&\leq CL\int_{t_0}^t e^{-(\gamma-\omega)(t-s)}(t-s)^{-\alpha}e^{-\gamma s}(\|u(s)\|_\alpha \\
&\quad + h(s))ds + M\|x\|_\alpha \\
&\leq CL\int_{t_0}^t e^{-(\gamma-\omega)(t-s)}(t-s)^{-\alpha}(S_t + H_t)ds \\
&\quad + M\|x\|_\alpha \\
&\leq M\|x\|_\alpha + CL\Gamma(1-\alpha)(\gamma-\omega)^{\alpha-1}(S_t + H_t)
\end{aligned}$$

Then, for $t \geq t_0$ this implies that

$$S_t \leq M\|x\|_\alpha + c(\gamma)(S_t + H_t) \tag{4}$$

Since $c(\gamma) < 1$, (4) implies that for $t \geq t_0$

$$S_t \leq (1-c(\gamma))^{-1}(M\|x\|_\alpha + c(\gamma)H_t)$$

which in turn implies that for $t \geq t_0$

$$\|u(t)\|_\alpha \leq e^{\gamma t}(1-c(\gamma))^{-1}(M\|x\|_\alpha + c(\gamma)H_t) \tag{5}$$

Suppose that $h(t)$ is bounded on $[t_0,\infty)$. Then for $t \geq t_0$, $e^{\gamma t}H_t = e^{\gamma t}\sup\{e^{-\gamma s}h(s) : t_0 \leq s \leq t\} \leq \sup\{h(s) : t_0 \leq s \leq t\}$ $\leq \sup\{h(s) : t_0 \leq s < \infty\}$. Thus, the right-side of (5) is bounded for $t \geq t_0$ and (i) is established. Suppose that $h(t) = 0(e^{\sigma t})$, where $\omega + (CL\Gamma(1-\alpha)^{1/1-\alpha}) < \sigma < 0$. Choose $\gamma = \sigma$ and observe that for $t \geq t_0$, $H_t \leq \sup\{e^{-\gamma s}Ke^{\sigma s} : t_0 \leq s \leq t\} = K$ where K is a constant independent of t. Thus, the right-side of (5) is $0(e^{\sigma t})$ and (ii) is established. Finally, suppose that $h(t) = o(1)$. Then, (5) implies that for $t_0 \leq t_1 \leq t$, $\|u(t)\|_\alpha \leq e^{\gamma t}(1-c(\gamma))^{-1}c(\gamma)\sup\{e^{-\gamma s}h(s) : t_0 \leq s \leq t_1\}$ $+ e^{\gamma y}(1-c(\gamma))^{-1}c(\gamma)\sup\{e^{-\gamma s}h(s) : t_1 \leq s \leq t\} +$ $e^{\gamma t}(1-c(\gamma))^{-1}M\|x\|_\alpha \leq e^{\gamma t}(1-c(\gamma))^{-1}M\|x\|_\alpha +$ $e^{\gamma(t-t_1)}(1-c(\gamma))^{-1}c(\gamma)\sup\{h(s) : t_0 \leq s \leq t_1\} +$ $(1-c(\gamma))^{-1}c(\gamma)\sup\{h(s) : t_1 < s < t\}$. This implies that $\|u(t)\|_\alpha = o(1)$ and (iii) is established.

In conclusion we remark that our results relate to the development given in [1] and, in particular, provide a clarification of Theorem 16.7 of [1].

REFERENCES

1. A. Friedman, Partial Differential Equations, Holt, Rinehart, and Winston, New York, 1969.

2. D. Henry, Geometric Theory of Nonlinear Parabolic Equations, to appear.

3. T. Kato, Perturbation Theory for Linear Operators, Springer-Verlag, New York, 1966.

4. A. Pazy, A class of semi-linear equations of evolution, to appear.

5. G. Webb, Exponential representation of solutions to an abstract semi-linear differential equation, to appear.

6. K. Yosida, Functional Analysis, Springer-Verlag, New York, 1968.

INDEX